# QUICK-TURNAROUND ASIC DESIGN IN VHDL

## Core-Based Behavioral Synthesis

## THE KLUWER INTERNATIONAL SERIES IN ENGINEERING AND COMPUTER SCIENCE

## VLSI, COMPUTER ARCHITECTURE AND DIGITAL SIGNAL PROCESSING

*Consulting Editor*

**Jonathan Allen**

### *Other books in the series:*

**ADVANCED CONCEPTS IN ADAPTIVE SIGNAL PROCESSING**
W. Kenneth Jenkins, Andrew W. Hull, Jeffrey C. Strait
ISBN: 0-7923-9740-1

**SOFTWARE SYNTHESIS FROM DATAFLOW GRAPHS**
Shuvra S. Bhattacharyya, Praveen K. Murthy, Edward A. Lee
ISBN: 0-7923-9722-3

**AUTOMATIC SPEECH AND SPEAKER RECOGNITION: Advanced Topics,**
Chin-Hui Lee, Kuldip K. Paliwal
ISBN: 0-7923-9706-1

**BINARY DECISION DIAGRAMS AND APPLICATIONS FOR VLSI CAD**, Shin-ichi Minato
ISBN: 0-7923-9652-9

**ROBUSTNESS IN AUTOMATIC SPEECH RECOGNITION**, Jean-Claude Junqua, Jean-Paul Haton
ISBN: 0-7923-9646-4

**HIGH-PERFORMANCE DIGITAL VLSI CIRCUIT DESIGN**, Richard X. Gu, Khaled M. Sharaf, Mohamed I. Elmasry
ISBN: 0-7923-9641-3

**LOW POWER DESIGN METHODOLOGIES**, Jan M. Rabaey, Massoud Pedram
ISBN: 0-7923-9630-8

**MODERN METHODS OF SPEECH PROCESSING**, Ravi P. Ramachandran
ISBN: 0-7923-9607-3

**LOGIC SYNTHESIS FOR FIELD-PROGRAMMABLE GATE ARRAYS**, Rajeev Murgai, Robert K. Brayton
ISBN: 0-7923-9596-4

**CODE GENERATION FOR EMBEDDED PROCESSORS**, P. Marwedel, G. Goossens
ISBN: 0-7923-9577-8

**DIGITAL TIMING MACROMODELING FOR VLSI DESIGN VERIFICATION**, Jeong-Taek Kong, David Overhauser
ISBN: 0-7923-9580-8

**DIGIT-SERIAL COMPUTATION**, Richard Hartley, Keshab K. Parhi
ISBN: 0-7923-9573-5

**FORMAL SEMANTICS FOR VHDL**, Carlos Delgado Kloos, Peter T. Breuer
ISBN: 0-7923-9552-2

**ON OPTIMAL INTERCONNECTIONS FOR VLSI**, Andrew B. Kahng, Gabriel Robins
ISBN: 0-7923-9483-6

**SIMULATION TECHNIQUES AND SOLUTIONS FOR MIXED-SIGNAL COUPLING IN INTEGRATED CIRCUITS**, Nishath K. Verghese, Timothy J. Schmerbeck, David J. Allstot
ISBN: 0-7923-9544-1

# QUICK-TURNAROUND ASIC DESIGN IN VHDL

## Core-Based Behavioral Synthesis

**Mohamed S. Ben Romdhane**
*Rockwell International Corporation*

■

**Vijay K. Madisetti**
*Georgia Institute of Technology*
*& VP Technologies*

■

**John W. Hines**
*US Air Force Wright Laboratories*

KLUWER ACADEMIC PUBLISHERS
Boston/London/Dordrecht

**Distributors for North America:**
Kluwer Academic Publishers
101 Philip Drive
Assinippi Park
Norwell, Massachusetts 02061 USA

**Distributors for all other countries:**
Kluwer Academic Publishers Group
Distribution Centre
Post Office Box 322
3300 AH Dordrecht, THE NETHERLANDS

**Library of Congress Cataloging-in-Publication Data**

A C.I.P. Catalogue record for this book is available
from the Library of Congress.

ISBN-13: 978-1-4612-8612-7 e-ISBN-13: 978-1-4613-1411-0
DOI: 10.1007/978-1-4613-1411-0



Softcover reprint of the hardcover 1st edition 1996



*Printed on acid-free paper.*

# CONTENTS

# LIST OF FIGURES

**Chapter 4**

**Chapter 5**

**Chapter 6**

**Chapter 7**

**Appendix A**

# LIST OF TABLES

# FOREWORD

Modern digital signal processing applications provide a large challenge to the system designer. Algorithms are becoming increasingly complex, and yet they must be realized with tight performance constraints. Nevertheless, these DSP algorithms are often built from many constituent canonical subtasks (e.g., IIR and FIR filters, FFTs.) that can be reused in other subtasks. Design is then a problem of composing these core entities into a cohesive whole to provide both the intended functionality and the required performance.

In order to organize the design process, there have been two major approaches. The top-down approach starts with an abstract, concise, functional description which can be quickly generated. On the other hand, the bottom-up approach starts from a detailed low-level design where performance can be directly assessed, but where the requisite design and interface detail take a long time to generate. In this book, the authors show a way to effectively resolve this tension by retaining the high-level conciseness of VHDL while parameterizing it to get good fit to specific applications through reuse of core library components. Since they build on a pre-designed set of core elements, accurate area, speed, and power estimates can be percolated to high-level design routines which explore the design space. Results are impressive, and the cost model provided will prove to be very useful. Overall, the authors have provided an up to date approach, doing a good job at getting performance out of high-level design.

The methodology provided makes good use of extant design tools, and is realistic in terms of the industrial design process. The approach is interesting in its own right, but is also of direct utility, and it will give the existing DSP CAD tools a highly competitive alternative. The techniques described have been developed within ARPA's RASSP (Rapid Prototyping of Application Specific Signal Processors) project, and should be of great interest there, as well as to many industrial designers.

Looking to the future, this design approach relies on the development of an interoperable and verified core library, as well as industry-wide standards so that the core elements can be effectively used in a wide variety of design environments. These achievements will greatly facilitate reuse of the core library designs, and spur the design exploration of high-performance systems to gain the high performance needed for contemporary complex signal processing tasks.

Prof. J. Allen
Director, Research Laboratory for Electronics,
Massachusetts Institute of Technology (MIT).

# PREFACE

This monograph describes a novel core-based design methodology for application-specific standard products (ASSP) and application-specific integrated circuits (ASICs) that promotes Design for Reuse (DfR) and Design with Reuse (DwR). While low level module-based design approaches have been pursued in the past, we define cores in our methodology as interoperable application-specific *and* organization-specific units of behavior (that can be synthesized into hardware) designed "off cycle". We propose that a core-based ASIC design methodology results in design cycles that are relatively short, and also allows rapid exploration of the design space, improving the quality of the result. In essence, it is a small beginning towards a Matlab$^{TM}$-like approach to hardware synthesis.

ASSPs appear to be the product of choice, with their share expected to reach nearly one half of the $10 billion digital signal processing (DSP) market by the year 2000. We propose and develop a methodology that enables organizations to establish an electronics system design environment that puts them at a competitive advantage through the creation of interoperable application-specific and organization-specific DfR libraries of cores described in VHDL. VHDL provides numerous advantages to the design methodology that will be highlighted through out this monograph, beside being a widely accepted and well documented standard in the commercial and the military industry. Thus, the methodology described here can be implemented within any standard EDA environment (e.g, Synopsys, Mentor, Compass, to name a few) at an organization consistent with concurrent engineering (CE) guidelines.

The reuse core library-based methodology for ASSP design was initially proposed by Madisetti in early 1993 during the proposal phase of US Department of Defense's DARPA's rapid prototyping initiatives (e.g., RASSP, ASEM). After a RASSP Technology Base award to Georgia Tech' DSP Laborotory, Ben Romdhane and Madisetti formally developed the DfR/DwR methodology as presented in this monograph, followed by benchmarking and validation on a number of ASSPs and ASICs, which culminated in Ben Romdhane's Ph.D dissertation at Georgia Tech in 1995; with Hines of US Air Force Wright Laboratories providing vital insight into the effective use of VHDL. Other independent efforts in this area are also documented within the monograph. A business case for this approach is also established to illustrate the suitability to both commercial and military markets.

The research documented in this monograph was supported in part by Defense Advanced Research Projects Agency (DARPA/ETO) of the US Department of Defense as part of the Rapid Prototyping of Application Specific Signal Processors (RASSP) Technology Base program (M. Richards and R. Harr) from 1994-1997, by the Government of Tunisia through a graduate fellowship to Ben Romdhane, by National Semiconductor Corporation (C. Shenhav and R. Batruni), by National Science Foundation (J. Cozzens), by Lockheed Martin Sanders (Bill Hood and Cory Myers), and in part by COMPASS Design Automation (C. Newhouse).

We thank our sponsors, colleagues, friends, and families for encouraging the writing of the monograph. We thank Mr. Carl Harris of Kluwer for his tireless efforts in bringing this book to publication, and to Prof. Jonathan Allen of MIT for generously writing the foreword.

M. S. Ben Romdhane
V. K. Madisetti
J. W. Hines.

# 1

# INTRODUCTION

## 1.1 PROBLEM DESCRIPTION

This monograph proposes the rapid design synthesis of Application-Specific Integrated Circuits (ASICs) for Digital Signal Processing (DSP) from a given set of design specifications through the use of library of functional cores. ASICs are the method of choice in DSP, when very high sample rates are sought in combination with low power and area requirements. Their drawback in comparison with programmable DSPs has been their long design times (12-18 months), lack of easy upgrades for legacy ASICs, very little hardware design reuse via libraries of DSP modules, and little capability to combine algorithmic design with lower level implementation tradeoffs. The traditional approach to design ASICs has always been through the following iterative procedure:

1. Customer requirements are documented.
2. Vendor converts requirements to specifications.
3. Vendor "trades off" a variety of possible implementations, selected manually or in an automated manner, that meet the specifications via simulation.
4. Vendor synthesizes acceptable design, verifies functionality and timing, and reiterates through the design process if specifications are not met.

The traditional full-custom handcrafted approach has been unable to meet the tight time-to-market requirements of recent IC products, while suitable for typical product schedules of the eighties. Several techniques have been proposed [1, 2, 3, 4, 5, 6, 7, 8, 9, 10, 11, 12] in recent literature to improve upon the basic procedure described above. With a shrinking time-to-market and increasing design complexity, it is critical that an organization make an appropriate choice of the design methodology. The goal is to meet the deadline and the customer's requirements while still conserving a cost/performance competitive advantage.

Most new approaches to ASIC design attempt to improve on the iterative design procedure by introducing higher-level design specification techniques primarily based on hierarchy and concurrency, and the use of automation, to assist with each step of the process. Other approaches propose a design methodology that allows the designer to guide the high-level design tool (interactively) to converge to an "acceptable-performance" design. Such approaches attempt to improve upon the efficiency by constraining either the domain of application, the size of the problem, or the flexibility in the architecture to be synthesized (e.g., Mentor Graphics' Mistral 2).

Still, problems such as extremely slow synthesis runtimes, inability to estimate (accurately) the cost function (e.g., in terms of area, delay, and power dissipation), early in the design process, problems with deep submicron design, issues of routability, limit the feasibility of architectural exploration and constrain synthesis to only low complexity designs [3, 13, 14] even among the most recent ESDA offerings from leading vendors.

Top-down design methodologies are efficient since they start at a higher level of abstraction, and also specify the interfaces between various components during the initial stages of the design. The problems with top-down design arise chiefly because the building blocks are often at a low level (e.g., gates), resulting in unsynthesizable design specifications due to lack of efficient global optimization algorithms. Bottom-up design methodologies capitalize on the systematic reuse of libraries at various levels, but the interfaces between the components of the libraries must be standardized to ensure correct and efficient implementation.

In order to implement both top-down and bottom-up design methodologies, methods that percolate exact area, timing, and power dissipation information to the initial phases of the design process are needed. Accurate design cost information of components is then known to the synthesis tool beforehand, and appropriate design decisions can be made at early stages of the design process with a very few variables to optimize over. Synthesis run times are also dramatically reduced since the netlist consists of pre-designed structural or behavioral cores instead of single gates.

For the purposes of this monograph, a *core* is defined as an application-specific macrocell (specified in a form that can be synthesized) that implements a unit of application-specific behavior. For instance, the core could implement a single butterfly in the FFT, or a single tap for a FIR filter. Alternatively, the core could be a synthesizable description of a programmable DSP that could be synthesized together with additional glue logic for peripherals, I/O, and memory to create an Application-Specific Standard Product . *This monograph proposes that systematic reuse of hardware and software cores and macrocells is the quickest and the most efficient approach to design application-specific and customer-specific ASICs. The monograph investigates how reuse can be harnessed efficiently within a VHDL-based environment to synthesize high-quality ASICs quickly and economically.*

This approach has been very successful in application-specific software synthesis using tools such as Matlab$^{TM}$. Matlab provides a variety of application specific tool boxes that allow the designer to rapidly create software executables. Hardware synthesis, while similar to software synthesis, has to optimize additional performance measures such as speed, area, and power. Thus, the applicability of the reuse-based approach to hardware synthesis has hitherto been an open issue, and is the focus of this monograph.

## 1.2 DESIGN MATURITY LAYERS

A design environment maturity diagram that is proposed in this monograph can be summarized by Figure 1.1. The three layers that correspond to the progressive maturity levels of the design environment of an organization are shown.

Most vendor organizations (and designers) have at their disposal a number of schematic capture tools, VHDL simulators, logic synthesis and FPGA synthesis tools, etc. These tools are basic necessities for any design task, and since they are available at all organizations, they form the Layer 1 environment whose maturity is at the level of *competitive necessity* [102] . Reuse of hardware is at a very low level, and typically involves the use of gate-level primitives.

At the next level, some organizations have tools and work flow environments that are customized to one or more target applications (e.g., DSP) of interest. These tools range from sophisticated application design and debugging capabilities in addition to a more complete set of synthesis tools, say, COM-

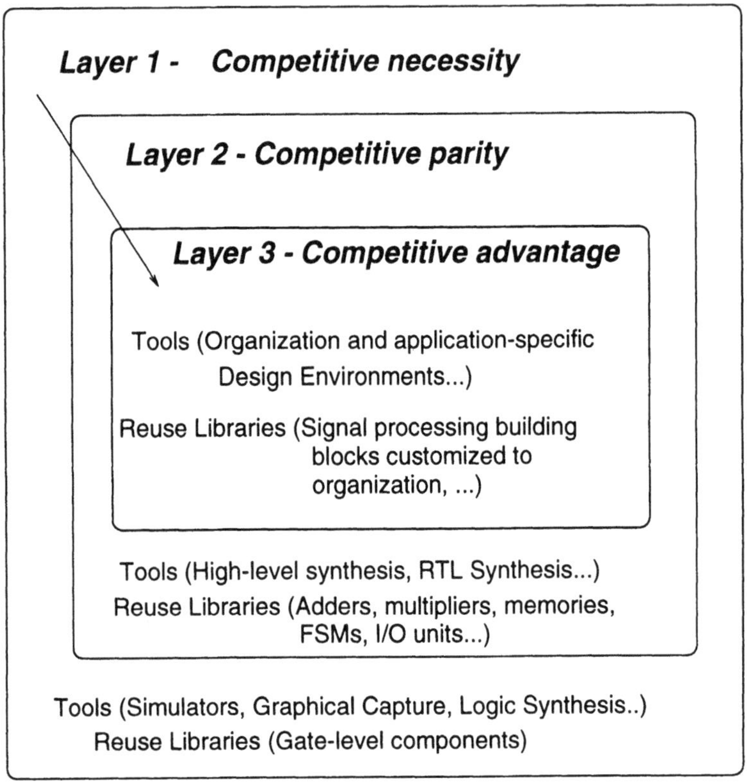


**Figure 1.1** *Design Environment Maturity Diagram.*

PASS's Datapath Compilers$^{TM}$, Mentor's DSP Station$^{TM}$, Synopsys' Behavioral Compiler$^{TM}$, or UC Berkeley's HYPER/LAGER environments. The reuse HW libraries at this level usually contain components such as adders, multipliers, and memory elements. The ASIC design environment uses these generic component libraries to synthesize circuits that are suited to a particular application. Since most organizations have access to such tools, we term this stage of maturity at Layer 2 as one of *competitive parity*. A few leading industries are in the process of developing reusable libraries at this level, primarily for software design, and more recently for hardware design.

The next level of design maturity is attained when the organization customizes the application-specific Layer 2 towards its own cost function and goals. The reuse libraries are application-specific and organization-specific building blocks (or functional cores) at a higher level of abstraction, akin to functional pieces of the target application, e.g., FFT butterflies, digital filter (FIR) taps, etc. It would not be economic to provide functional cores at Layer 2 since the number of choices (when not considering the organization-specific needs and objective functions) can be very large (see discussion of sizing the component libraries in chapter 4). Thus, Layer 3 provides a design environment that leads to a *competitive advantage* for an organization.

## 1.2.1 Support for reuse

Synthesis tools at Layer 2, such as the Behavioral Compiler$^{TM}$ from Synopsys, map a subset of VHDL onto a pre-defined datapath, controller and memory architecture taking into consideration the area and latency constraints specified by the user. VHDL processes are mapped onto a DesignWare$^{TM}$ library based on multipliers, adders, memories, controllers and I/O support. Our approach instead uses a library of pre-synthesized (off-cycle, possibly with the Behavioral Compiler and integrated into the DesignWare library) application-specific cores (efficiently floorplanned and routed) that are reused to create new application-specific chips. It is expected that these cores are parametrizable and customizable to synthesize a larger application specification very quickly. Of course, not all applications can have pre-fabricated cores available, and a responsibility of the design organization is to predict future implementation needs and populate an extensive library of pre-designed blocks. For instance, a core could be based on the internals of a DSP that can be customized as an ASIC core and utilized within the ASIC design environment. With the onset of deep submicron designs, such pre-designed components would be accurately characterized off-cycle for simulation and test purposes reducing the risk associated with implementation. Ironically, this is a structural approach to rapid behavioral synthesis, wherein the target application is structurally partitioned into a number of reusable pre-designed cores. Indeed, Synopsys also suggests that multiple entity designs be generated via partitioning for very complex ASICs, especially when the number of operations to be scheduled per entity is over 150. *Thus our methodology can be implemented via most existing RTL and behavioral synthesis tools based on VHDL as is shown in the rest of the monograph.* Our choice of COMPASS ASIC Synthesizer is coincidental, and does not represent the only choice available to implement the maturity model-driven methodology for ASIC design.

We propose that this layered approach to ASIC design leads to a very competitive quick-turnaround ASIC design environment that is customized to the organization and application requirements and constraints. A low risk and highly profitable design environment can thus be developed by a candidate organization wishing to use the approach presented in this monograph. As mentioned earlier, software developers may see the similarity of the proposed hardware synthesis approach to the Matlab$^{TM}$ approach to DSP software synthesis through the use of pre-coded software macros in a signal processing toolbox.

## 1.3 PROPOSED APPROACH

The design methodology for ASICs proposed by this monograph is based on systematic functional hardware reuse that is customized to application and the organization. The components reused are not adders, multipliers and memory elements typically utilized in high-level design tools, but functional cores (consisting of at least 5K+ gates) or macros for a class of applications (e.g., FIR filter taps, FFT butterflies, etc) that suit organizational objective functions (e.g., functional blocks utilized by a commercial vendor, such as National Semiconductor Corporation, could be optimized for area and power, while those of a military vendor such as Lockheed Martin could be primarily optimized for speed, testability, and/or radiation-hardness). Reuse that is considered at the functional level provides higher payoff in terms of ease-of-use and design turnaround because it takes the designer much less work to assemble, optimize, and verify the design using large-scale components. Yet, high performance and tremendous savings can follow from the use of highly pre-optimized and pre-verified reusable large-scale components (functions).

Figure 1.2 summarizes the proposed approach. An expert team of library developers utilizes both application-specific and organization-specific constraints to provide a number of reusable and interoperable (synthesizable and simulatable) cores during the Design for Reuse (DfR) phase that is carried out "off-cycle". The designers specify the ASIC using these cores, perform performance tradeoffs, timing and functional verification, before synthesis. This Design with Reuse (DwR) phase is completed "in cycle". Two approaches are proposed during the ASIC specification phase. The first, an Automated Design Line (ADL) , requires that only a few functional and performance requirements of the ASIC be given, and the environment uses previously prepared configuration file to synthesize the ASIC in minutes by customizing constituent cores.

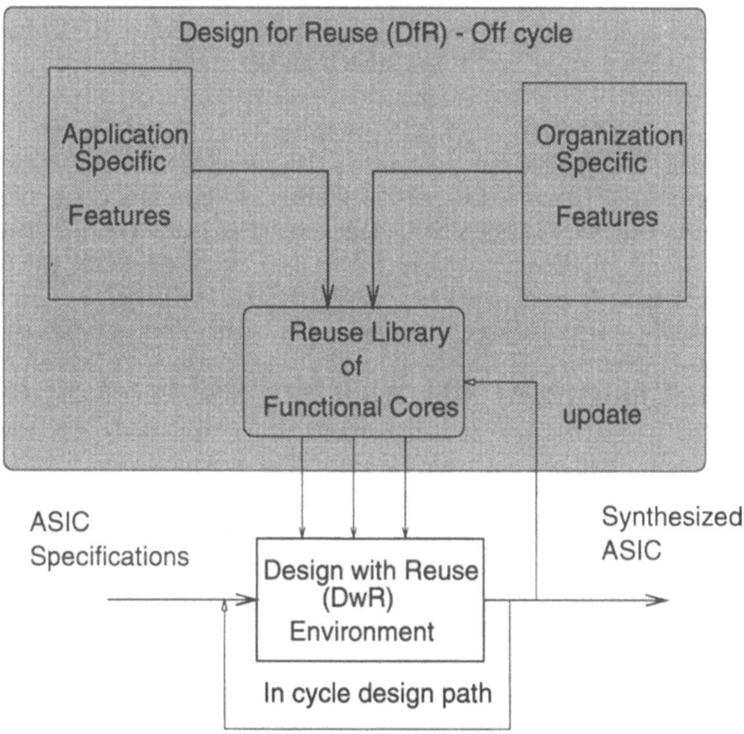


**Figure 1.2** *Proposed Core-Based Design Environment.*

This is akin to designing a FIR filter by specifying the number of coefficients, the clock rate, and constraints on area and power dissipation, without any knowledge of any hardware description language (HDL), such as VHDL. The second approach, called *Functional Design Line (FDL)* , requires specification of the ASIC in terms of composition, generalization, and interconnection of the various building blocks and cores in the reuse library, and requires knowledge of VHDL by the DwR design team to implement the glue logic and interfaces.

Efficient functional reusability is derived from the built-in flexibility of parameterized design with VHDL [15], and its many flexible constructs and characteristics (i.e., abstraction, encapsulation, inheritance, and reuse through *types*, *attributes*, *packages*, *procedures*, and *functions*). Efficient and rapid design exploration is realized through a limited but realistic design space exploration bounded by the proposed reusable datapath library. Component identification, classification, storage, and retrieval, within the proposed environment, are handled systematically. A Tcl/Tk graphical user interface (GUI) has been developed for his purpose and is available from the authors.

The reuse-based environment broadly considers three design views—(1) The *core or module* design view corresponds to a Design-for-Reuse (DfR) cycle. This is the off-cycle generation of the reusable library of functional cores. The core design view answers questions that include, *how does one identify a functional or datapath core for the reuse library ?*, *how does one increase the potential of reuse of such cores ?*, and *how does one model the cores in VHDL ?*. This view is undertaken off-cycle and off the critical path, and hence does not directly affect the time-to-market cycle of ASICs. (2) The *system* design view corresponds to a Design-with-Reuse (DwR) cycle. This is the cycle where reusable cores and modules (both synthesizable and simulatable) from a pre-designed library are used to design, optimize, and verify ASICs. The system design view answers questions that include, *how does one automate the design process ?*, and *how does one explore the design space ?*. This DwR is undertaken in-cycle. (3) The *simulation* design view is related to the validation of the functionality of the design and the conformity of its performance to the original design specifications and is primarily the verification phase.

A detailed cost and time model is proposed to estimate the cost and design turnaround of this reuse approach, and to compare the proposed methodology with the conventional design of ASICs and Field Programmable Gate Arrays (FPGAs).

This approach was utilized to rapidly describe and design digital filters and FFT processors (in few person-days) with the capability to explore a variety of design alternatives in very short times. The resulting ASICs are comparable in performance with the best implementations reported in recent literature as will be discussed chapter 5.

## 1.4 MARKET TRENDS

Studies by market research organizations such as Forward Concepts show that there is a growth in DSP core-based ASICs, also called Application-Specific Standard Products (ASSPs) and a decline in the sales of general purpose DSPs. Core-based ASICs constituted about 35% of the nearly $2 billion market in DSPs in 1995, and it is projected that their share will grow to 47% of the

$10 billion market for year 2000. Thus core-based ASICs will compete with custom ASICs and general purpose DSPs in the near future. General purpose DSPs expect their market share to drop from 50% in 1995 to 26% in 2000. The remainder of the market is taken by custom ASIC solutions which are projected to be 27% of the market in 2000.

More vendors are increasingly dependent on ASSPs to meet cost and performance objectives, while maintaining short times-to-market in the current global market place. Semiconductor vendors such as Analog Devices, Motorola, and Texas Instruments are making available many of their processors as cores for incorporation into ASSPs. Other companies and organizations such as VLSI Technology, Boeing, DQDT, Aspec, Department of Defense, NEC, DSP Group, and Mitsubishi are also creating customizable reusable core libraries for rapid ASIC design in the area of DSP. To support this new design methodology and to provide design support for system level integration, detailed hardware/software co-simulation is necessary. This requires that software execute on cycle accurate models of the hardware. As part of the ARPA's Rapid Prototyping of Application-Specific Signal Processors (RASSP) program, Georgia Tech has developed a number of cycle-accurate models for a variety of DSP cores and processors that have been used industry wide. Companies such as Synopsys, Cadence, Mentor Graphics, Sanders, Lockheed Martin, Hughes, Honeywell, VP Technologies, and Zuken, are developing modeling environments for system-level design, and HW/SW integration and test through systematic use of VHDL models at multiple levels of design abstraction.

## 1.5 ORGANIZATION OF THE MONOGRAPH

This monograph starts by reviewing some background material in Chapter 2. First, existing ASIC design tools and approaches are introduced, and their advantages and limitations are highlighted. These approaches are then classified based on their ease-of-use and design flexibility.

Chapter 3 describes the motivation behind the use of VHDL and introduces the parameterized design with VHDL as the basis for introducing flexibility into the design. This chapter also describes the constructs of the language that provide the required flexibility to define an efficient and systematic reuse-based design approach.

Chapter 4 introduces our proposed functional reuse-based design methodology and environment. The different design views (*module, system, simulation*) that constitute the proposed approach are discussed in detail. These design views include issues such as — development of a reusable library (design, flexibility, and ease-of-use), systematic component identification, classification, storage, and retrieval, and verification issues. The proposed reuse approach is compared with other existing design approaches.

Chapter 5 describes the rapid prototyping of typical DSP applications within the proposed environment. Such applications include the prototyping of digital filters and FFT processors.

Chapter 6 discusses the issues of board integration and describes how to validate the functionality of the designed ASICs in the system.

Finally, Chapter 7 summarizes the contribution of the monograph and discusses future research issues.

# 2

# BACKGROUND

This chapter reviews the background required for this monograph. First, common ASIC design approaches are introduced. Advantages and limitations of ASICs with regard to general-purpose DSP processors are outlined. Section 2.2 describes popular design approaches and environments proposed in recent literature to synthesize ASICs form HDL descriptions. In section 2.3, a comparison between existing ASIC design tools is proposed based on criteria such as — design turnaround, flexibility, ease-of-use, and performance. Section 2.4 introduces a cost model originally proposed by Synopsys and Xilinx that compares the ASIC and FPGA alternatives. This cost model and assumptions made will serve us later, when we compare our proposed design methodology with other high-level approaches.

## 2.1 THE ROLE OF ASICS

ASICs have traditionally provided large-volume low-cost solutions to a variety of application needs. Concurrently, general purpose processors and specialized digital signal processors have continued to develop and improve offloading many applications from the exclusive domain of ASICs. With the advent of the multimedia age, and the requirement for real-time, low power, and high performance required, ASICs have been utilized to implement increasing complex functions (from video conferencing systems, mobile phones, to automotive controllers) thus greatly extending their lifespan as a method of choice for implementation.

The performance of general-purpose Digital Signal Processors (DSPs) [19], though far superior to general purpose microprocessors for signal processing applications, has been limited (due to the need for programmability) by the instruction fetch and decode that precedes the execution of an operation, introducing an additional overhead to every execution clock cycle. This overhead in speed is diminished through the use of pipelining, though the overhead in area and power due to the need for address generators, memories, and multiple busses can be significant. Users can write high quality code that optimizes the execution by means of techniques such as software pipelining and instruction pre-fetching. DSPs, however, will continue to increase in flexibility, ease of use, and provide computational capabilities (in the multiprocessor context) at a bargain price in the next few years.

ASICs and ASSPs have some performance advantages over digital signal processors (DSPs) in speed, size and power, but lack the programmability and ease of portability. The speed, size, weight, power advantages are gained when area is optimized by eliminating unnecessary multiplexors and all unwanted functions that are part of a standard DSP. ASICs drive less capacitive load, resulting in less power dissipation and better performance. The word precision and other design parameters can be tailored to an application to provide optimum high-level performance. A better design reliability is obtained as a result of a single-chip implementation that eliminates some of the external interconnects. The flexibility of the design to meet the speed, area, and power constraints is obtained by tagging the design to the appropriate hardware resources. In addition, the advent of advanced multichip (MCM) packaging technology enables the use of DSPs, ASSPs, and ASICs within the same implementation, thus ensuring that the system designer can take advantage of both these implementation alternatives. It is expected that hardware and software implementation of the future will contain DSP cores, microprocessors, ASICs, ASSPs, and FPGAs based on specific customer-specific and organization-specific objective functions.

## 2.2 ASIC AND ASSP DESIGN APPROACHES

The current practice ASIC and ASSP design procedures followed in most commercial designs was discussed in Chapter 1. High-level synthesis, loosely described, attempts to solve design problems at a conceptual level in the design space, freeing the designer from dealing with complex, numerous and intricate low level details.

This design space can be represented as a triplet:

*(Function, Performance, Constraints)*

The ASSP has to deliver a function $F$ rated at performance $P$ (sample rate, MIPS, etc) under some constraints $C$ (area, power, etc). From the perspective of high level synthesis, the function $F$ is considered at the algorithmic level (a behavioral HDL code is written to describe the algorithm). The performance $P$ is obtained by undertaking different high level optimization techniques such as—retiming, applying algebraic laws, and loop unrolling and shrinking [19, 20]. The constraints $C$ are satisfied by tagging the final design to the appropriate hardware resources (i.e., portability suggests low power; military environment suggests high temperature).

Before its actual implementation into hardware, a design ripples through a sequence of refinement stages. These stages are also called abstraction levels. High level synthesis implements such a refinement process in an automatic way, following a **top-down** design methodology (see section 2.2.1). The different design abstraction levels of the top-down design methodology are illustrated by Figure 2.1 [17]. The highest abstraction level, the specification level , defines the functionality, the performance, and the constraints of target application either textually or graphically. At the behavioral level , the designer writes a software program that executes the function to be implemented in hardware. Only the high level performance of the algorithm is verified at this stage. At the register transfer level (RTL) , the behavioral description of the design is tagged to a well-defined hardware microarchitecture. RTL is a level of design abstraction in which the behavior and structure of a design is specified by describing the operations that are performed on data as it flows between circuit inputs, outputs, and clocked registers [2]. The logic level introduces further refinement into the design. The different blocks of the hardware architecture are now committed to Boolean logic gates, latches, and flip-flops (i.e., at this point an adder is committed to a specific implementation, for instance, a carry-ripple or a carry-save implementation). At the circuit level , the logic circuitry is transformed into transistors. The final design abstraction level, before sign-off and transmission of the chip to the foundry, is the layout level . Here the design is linked to a specific technology and mask geometries are derived.

High level synthesis is usually initiated at the RTL level [2, 21, 18] to ensure success. The synthesis of an ASIC from a behavioral description (behavioral synthesis) is also possible in theory. In practice, however, the very large number

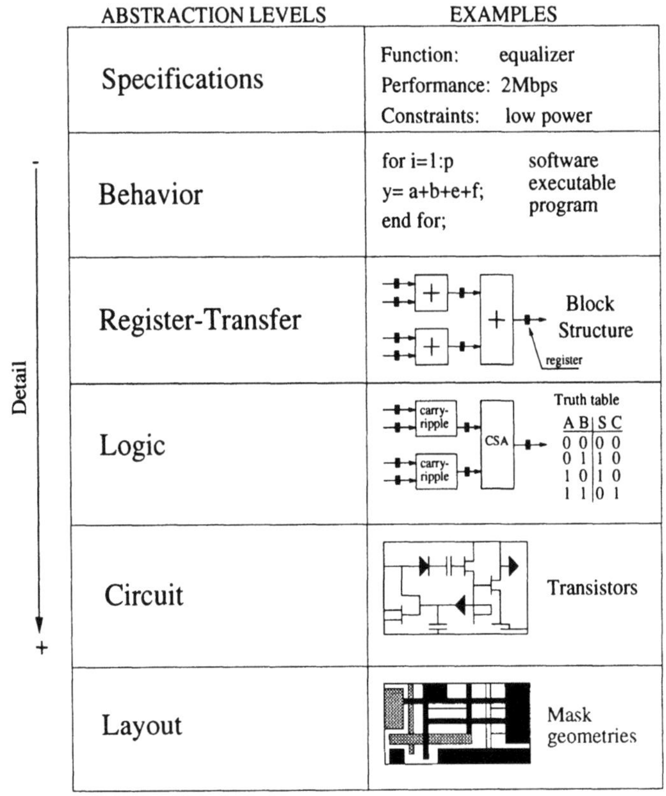


**Figure 2.1** *Design abstraction levels in a top-down design methodology.*

of variables in a complex optimization space results in suboptimal efficiencies. Some practical approaches compromise between design complexity and product performance, resulting in top-down semi-custom design approaches. Here, high-level synthesis tools are interactively assisted by designers throughout the entire design process. Such interaction helps a rapid and efficient convergence to the final IC design. Still, problems such as extremely slow synthesis runtimes (i.e., due to random logic synthesis (see Table 2.3)) and the inability to estimate accurately the cost function (i.e., in terms of area, delay, and power dissipation) early in the design process, limit the feasibility of architectural exploration and constrain synthesis to only low complexity designs [3, 13, 14].

As mentioned earlier, design methodologies that percolate exact area, timing, and power dissipation information to higher abstraction levels are needed.

These are design methodologies that require a library of pre-designed RTL components. Hence, accurate design cost information of components is known to the high-level synthesis tool beforehand, and appropriate design decisions can be made at early stages of the design process. Synthesis run times are also dramatically reduced since the netlist consists of pre-designed physical macrocells and cores instead of single gates. Datapath compilers have been proposed [22, 23, 24, 25] as bottom-up design tools. The datapath-inherent structure of the design is extracted and synthesized from a datapath library comprising highly optimized layout components such as adders, multipliers, ALUs and memory storage elements.

Field Programmable Gate Array (FPGA) based synthesis is another established practice for the rapid prototyping of hardware systems. The FPGA alternative is gaining popularity for its short time-to-market cycle and low cost in small to medium volumes. The approaches proposed in this monograph can also be used in the FPGA arena.

The remainder of this section describes the top-down, bottom-up, and hybrid approaches to ASIC design in relation to FPGA-based design methodologies. Their advantages and limitations are also highlighted.

### 2.2.1 Top-down design

A top-down design methodology starts the design at the top of its abstraction hierarchy (i.e., algorithmic level). Functional and timing detail is gradually added to the design as it approaches the physical implementation. Several steps must be carefully followed in order to produce an acceptable implementation. As illustrated by Figure 2.2 [26], the design is first described behaviorally using a hardware description language (HDL). VHDL is recommended in this monograph for its many benefits; easy interfacing, excellent documentation, and system level simulations prior to fabrication [27, 28, 21, 29, 30, 31, 32, 18, 33]. This behavioral representation is then transformed to a consistent format (or internal representation) that can be understood and easily manipulated by the high-level synthesis tool. A popular choice for the internal representation is the data flow graph (DFG) or the fully-specified flow graph (FSFG) [19] representation [34, 35, 36, 37, 38]. Such internal representation also resolves the syntactic-variation problem that results from the synthesis of semantically equivalent but syntactically-different descriptions [26]. The high-level synthesis tool then operates on the DFG representation through graph transformation, resources allocation, scheduling, and hardware binding [39, 19, 20, 40, 41, 42].

The DFG optimization is conducted with regard to the original design constraints and the available resources in a hardware library.

Top-down design makes key decisions at several abstraction levels in order to synthesize the design that best fits the application's requirements (speed, area, power dissipation). Unfortunately, these decisions are made with little available information, since it is difficult to predict accurately the area, speed, and power dissipation of the final product when it is synthesized from gate-level leaf cells. Despite this fact, most high-level synthesis tools still map (in a top-down manner) regular bit-slice datapath structures to random logic. This approach faces two serious problems [3]:

1. Run times are extremely slow, limiting the feasibility of architectural exploration and constraining the synthesis to only low-complexity designs (as described below).
2. The synthesized circuits (if routable) result in sub-optimal implementations due to the scattering of the datapath across layout. The inefficiency of implementation translates to all performance criteria (i.e., speed, area, power consumption, and clock skew).

These problems result from several causes [3, 13, 14]. Control logic maps well to random logic due to its inherent single-bit structure that makes every logic operation a unique synthesizable operation. On the other hand, datapath-intensive operations are different from control logic in the sense that behavioral arithmetic operations, such as:

$$S = A + B, \tag{2.1}$$

are first expanded into Boolean equation pairs at the bit level:

$$S_i = A_i\bar{B}_i\bar{C}_i + \bar{A}_iB_i\bar{C}_i + \bar{A}_i\bar{B}_iC_i + A_iB_iC_i, \tag{2.2}$$

$$C_{i+1} = A_iB_i + A_iC_i + B_iC_i, \tag{2.3}$$

where $i = 1, ..., N$; $N$ is the precision; $C_i$ is the carry in, and $C_{i+1}$ is the carry out. The Boolean equation pairs 2.2, and 2.3 are then optimized and mapped into logic gates. As the complexity of the datapath architecture increases, the number of variables involved in such an optimization procedure grows exponentially, resulting in interminable runtimes and detrimental implementation efficiencies [13]. Moreover, architecture exploration becomes infeasible. For example, the Boolean expansion of Equation 2.1, illustrated by Equations 2.2 and 2.3, amounts to a ripple-carry implementation of an adder. A faster hardware implementation (i.e., carry-lookahead adder) requires a new Boolean expansion of Equation 2.1 and a re-run of the optimization and mapping procedures (i.e., mapping to standard cells). Efficient VHDL-based synthesis further requires the right usage of hierarchy in the input specification. A very hierarchical design can split the critical path along many components, and a flat design can results in a rat-nest routing.

Current top-down synthesis attempts to improve efficiency by constraining either the domain of application, the size of the problem, or the flexibility of the architecture to be synthesized [21, 43, 44, 45]. Once off-the-record, synthesis vendors recommend breaking the design into chunks (strictly in neither flat nor hierarchical organization) of logic of between 4-5K gates each that can be synthesized explicitly, and the entire design is then easily routed and floorplanned with assistance from the designer.

Optimization and synthesis algorithms used in actual top-down design will require major improvements in the use of advanced optimization algorithms [19] before fully automated top-down synthesis from behavioral specifications can stand by itself as a proven design methodology for complex ASICs.

## 2.2.2 Bottom-up design

A bottom-up design methodology that relies on a pre-designed interoperable RTL-component library of datapath cores and user-defined macrocells can be quite efficient. Resources are allocated from the core and macrocell library and tagged to a specific architecture. The tradeoffs between cost and performance are easily drawn and the design exploration space is quickly narrowed to a manageable level. Bottom-up design decisions are made with exact cost information provided by the pre-designed RTL-component library. As illustrated by Figure 2.3, datapath and control are designed separately and synthesized from different libraries; the datapath is synthesized using the RTL-component library while the control is synthesized in random-logic using a standard-cell library. Behavioral operations that do not match cores in the RTL-component library are also synthesized in random-logic.

Datapath compilers , like those from COMPASS Design Automation, are typical bottom-up synthesis tools. They are offered by several CAD vendors [22, 23, 24, 25] as a remedy to the inefficient sub-optimal top-down synthesis (in ran-

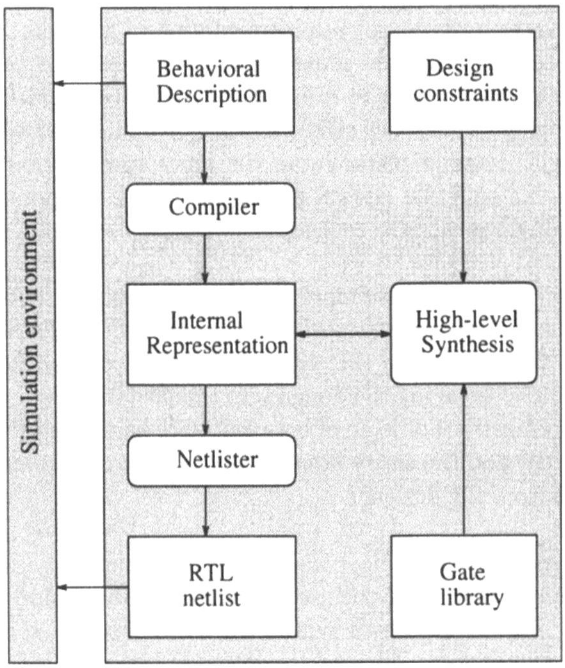


**Figure 2.2** *A Top-down design methodology.*

dom logic) of datapath-intensive structures. These tools take advantage of the regularity in the datapath specification; regular datapath bit-slices are mapped into pre-optimized datapath components [13, 23]. This procedure incurs far less synthesis runtime since the datapath core is assembled by simple abutment of datapath bit-slice blocks provided by a standby datapath library. The cells provided by the datapath library are optimized off-line. When no such tools are used (e.g., the datapath is implemented using classical standard cell place and route tools), the regularity of the datapath structure is lost during the implementation, resulting in a scattered layout implementation that adversely affects the area, speed, and power consumption of the synthesized design.

Arithmetic operations are typical datapath structures that are characterized by an inherent regularity that is optimal for datapath synthesis. Datapath compilers potentially yield substantial improvements in the following aspects:

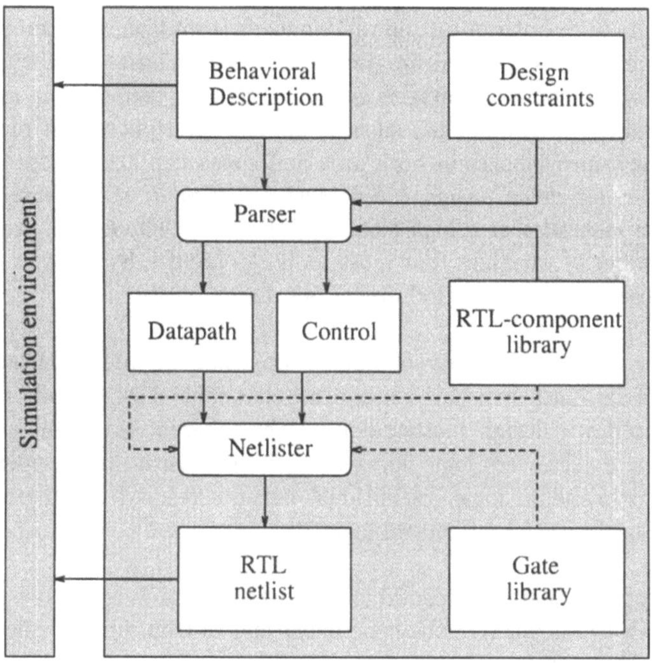


**Figure 2.3** *A Bottom-up design methodology.*

- *Speed* — The use of a datapath compiler yields a uniform clock delay that results in improved speed.
- *Area* — Datapath mapping is characterized by a regular structure that translates to efficiency at the layout level and results in optimized chip area.
- *Power* — Regularity at the layout level also translates into better power balance.
- *Optimality for rapid prototyping*— Datapath synthesis results in reduced synthesis runtimes. As a result, it provides the designer with more time to explore the design exploration space, and to find the best implementation. Datapath synthesis also allows flexibility in choosing different datapath libraries from several vendors optimized for different criteria (i.e., high speed, high density, low power, etc).

Table 2.3 illustrates design examples that have been synthesized[1] from a VHDL description in both bottom-up and random logic (top-down with gate-level components) methodologies. A dramatic reduction in synthesis run times can be noticed when bottom-up synthesis is used, primarily through the use of interoperable datapath components such as adders, multipliers and multiplexors. Remarkable improvements in both area and speed can also be obtained. *It is expected that top-down design will be just as efficient, if the library of reuse components were also at a high level of abstraction.* The key at present is the limited number of variables that most tools can handle; Synopsys's Behavioral Compiler can handle about 150 events per design entity.

In contrast to the suitability of bottom-up using high-level libraries to the synthesis of datapath structures, it appears that control logic is best synthesized using a top-down design methodology. The inherent single-bit structure of control logic makes every logic operation a unique synthesizable operation that maps well to random logic. Additional benefits result from the maturity of automatic optimization of control logic [2].

Thus, it appears that an acceptable synthesis approach is usually a combination of both top-down (for the controller) and bottom-up (for the datapath) synthesis approaches.

### 2.2.3 FPGA-based synthesis

Field programmable gate arrays (FPGAs) are a form of programmable logic designs that target the rapid prototyping of hardware systems at virtually no non-recurring engineering (NRE) cost [46, 47]. They have been proposed as a potential solution to the critical time-to-market pressures of the actual hardware design market. Their major advantage over conventional Programmable Logic Arrays (PLAs) is their full function controllability. That is, designers are no longer constrained by a traditional programmable logic device architecture, but instead can configure the device to any desirable logic function. Their advantage over the standard cell custom design is their low development cost and their very short design turnaround cycles.

[1] on a Sparc10 workstation

Table 2.1: Design synthesis ($0.8\mu$ CMOS) example of datapath structures†

| Design | bottom-up ≡ datapath synthesis | | | top-down ≡ random logic synthesis | | |
|---|---|---|---|---|---|---|
| 16-bit | Runtime | Area ($mil^2$) | Speed (MHz) | Runtime | Area ($mil^2$) | Speed (MHz) |
| 1-tap FIR | 0mn52s | 2654.12 | 27.9 | 7mn0s | 4421.54 | 20.5 |
| 4-tap FIR | 1mn03s | 10876.29 | 20.9 | 9h35mn25s | 14397.29 | 18.2 |
| 8-tap FIR | 1mn30s | 24944.28 | 19.2 | crashed! | - | - |

†*COMPASS Synthesizer is used*

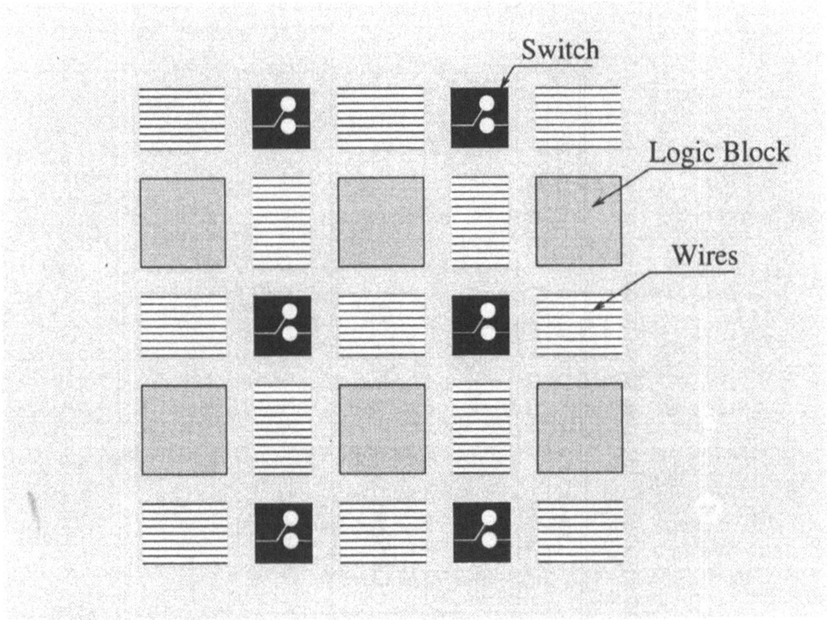


**Figure 2.4** *A typical layout structure of an FPGA.*

## *FPGA methodology*

A typical architecture of an FPGA is illustrated by Figure 2.4. It consists of — (1) uncommitted logic blocks, (2) programmable wires, and (3) programmable switches. The uncommitted logic blocks are disposed in an array structure that is interconnected through the programmable wires and switches. Full controllability over the functionality and the architecture of the target design can be obtained. The uncommitted logic blocks are either standard-function configurable logic blocks or programmable devices that can be programmed to perform any logic operation. SRAM cells and antifuses are the commonly used technologies for this purpose. The switches and wires are programmed to make the required interconnections between these logic blocks in order to synthesize the function of interest [47].

The design methodology for FPGAs has to take into consideration several characteristics (structure, advantages, limitations) that distinguish them from traditional standard cell ASIC design. Hence, special purpose design tools are required for FPGAs. The typical design process for an FPGA-based implementation is illustrated in Figure 2.5 [47].

FPGA design can be characterized by ( [47, 29]):

- Design entry – VHDL representation is recommended for reasons of portability maintainability, easy modification, debugging and reuse.
- Logic Synthesis – The major reason behind the short design turnaround of FPGAs is the fact that the long IC fabrication time and test for these devices is eliminated. By adopting logic synthesis, the front-end design cycle of FPGAs is also shortened. The output of logic synthesis is a set of Boolean expressions for combinatorial logic.
- Special purpose optimization – The Boolean equations generated by logic synthesis have to be transformed to optimize FPGA structure related constraints, (such as to minimize the number of FPGA logic blocks). FPGA-specific optimizations also have to be undertaken at all design levels. When the design is being physically mapped into an FPGA, geometrical floorplanning issues must be resolved to optimize parameters like the critical path or the total length of interconnecting wires. During the routing phase, issues such as 100% routability versus critical path minimization may have to be compromised. This issues of pinout to logic ratios also have to be considered.

## *Advantages and limitations*

FPGAs possess several important advantages. Their short design cycle and low development cost head this list. In comparison with other programmable logic devices, FPGAs have an added capability to customize designs to fit any logic function. Hence, designers are no longer constrained by a traditional programmable architecture. This added capability also allows for a dynamic reconfiguration of systems. This feature can be utilized to quickly explore and evaluate architectures for a target algorithm. For instance, The Splash-2 platform [48] is an illustration of the use of dynamic configuration of FPGAs. The platform is an array board composed of 16 fully-connected Xilinx XC4010 FPGAs that can be dynamically configured to quickly prototype and evaluate a variety of real-time image processing algorithms. Finally, being fully programmable, FPGAs are fully observable with no need for extra test logic.

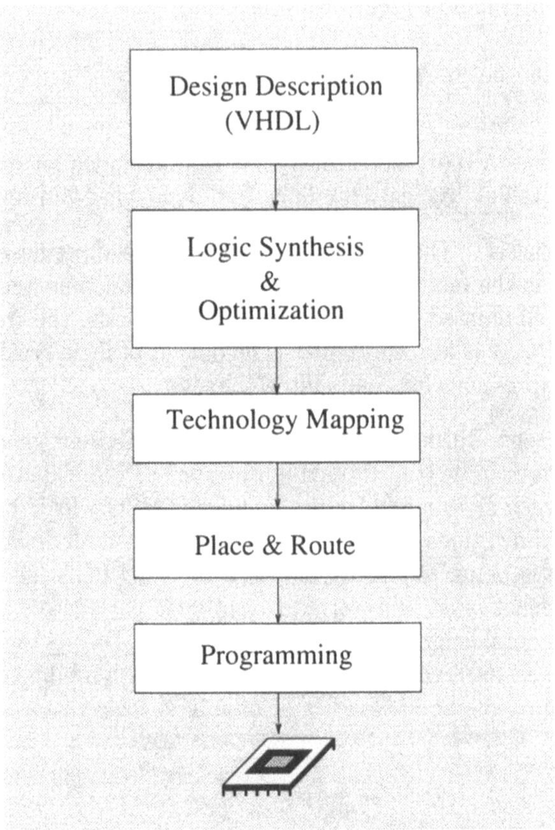


**Figure 2.5** *FPGA-based design.*

The advantages of FPGAs, listed above, come at the expense of some performance limitations. FPGAs have a serious speed and density limitation problem. FPGAs are only suitable for clock speeds around 10-25MHz at present. The programmable switches and wires accommodated by FPGAs (for routing purposes of the logic blocks) represent a large density overhead, while the utilization of uncommitted logic blocks can never attain 100%. Practical FPGA solutions target applications with an order of magnitude of 2000-5000 gates. FPGA prototyping of large applications requires multiple FPGA-prototype partitioning, which raises other design problems such as an even tighter speed limitation. The use of submicron technology to accommodate higher densities and data rates for FPGAs does not work well because of all the interconnections of an FPGA structure. Interconnect delay in submicron technology becomes more important than gate delay, hence it must be accounted for. Finally, because the programmability feature of FPGAs provides full device-observability, the functional testing of these devices is still handled in an ad-hoc manner.

## 2.3 CLASSIFICATION OF APPROACHES

The final goal of any design methodology is to provide users with desirable features such as — flexibility, ease-of-use, short design turnaround, high performance, integrability, accuracy, portability, and reliability.

Existent design methodologies and environments only favor a few of these desirable features. A classification of existing DSP environments based on ease-of-use and design flexibility is illustrated by Figure 2.6.

FIRGEN [49], for instance, follows a turn-key approach that provides ease-of-use but lacks flexibility and ease of integrability into larger systems. FIRGEN is dedicated to the rapid prototyping of fixed-coefficient FIR filters. It starts the design from the high level specification and takes it down to the physical implementation.The design cycle consists of three major steps: (1) high level filter design, (2) architecture and floor plan generation, and (3) layout generation. Some of the constraints of FIRGEN as reported are — lack of flexibility in the specification of optimization techniques, and lack of programmability of the coefficients, thought it would serve as an excellent standalone ADL environment.

Design environments that are based on symbolic layout and compaction techniques such as Mulga [5] provide better flexibility but less ease-of-use than a turn-key approach (such as FIRGEN). Designs can be retargeted to take advantage of different fabrication processes simply by recompiling the original design using a new technology file that corresponds to the desired process. Philips [50], Bellcore [5], and AT&T Bell Labs [51], are technology leading companies that have adopted symbolic layout within their design cycle in order to provide such flexibility.

Behavioral synthesis (top-down) through Silage/DFL [52, 40] reduces the number of details involved in IC design and provides a fair amount of ease-of-use and flexibility of application. However, it is still not mature enough to handle efficiently high complexity designs. Design tools that adopt this approach

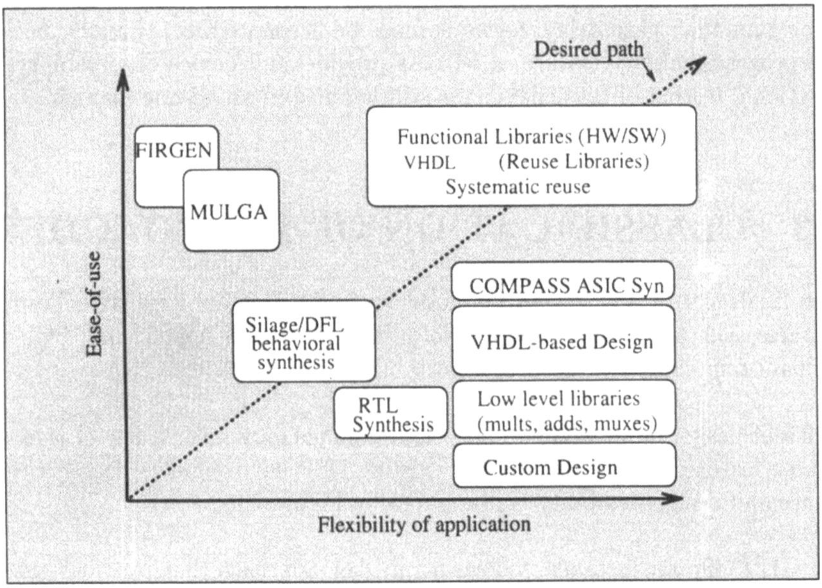


**Figure 2.6** *Classification of various DSP design environments.*

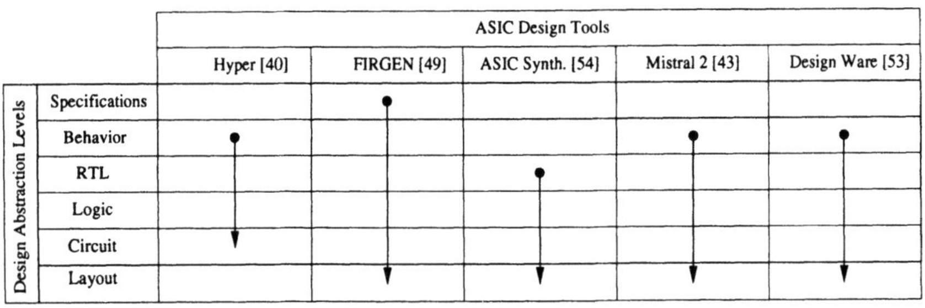


**Figure 2.7** *Design abstraction levels covered by existing DSP design environments.*

attempt to improve upon the performance of synthesized circuits by constraining the flexibility in the synthesized architecture through the use synthesis directives (i.e., the use of pragmas in Mentor graphics' Mistral 2 [43]), or by compromising between design complexity and product performance. They result in top-down semi-custom design approaches that are interactively assisted by designers throughout the entire design process. Such interaction helps a

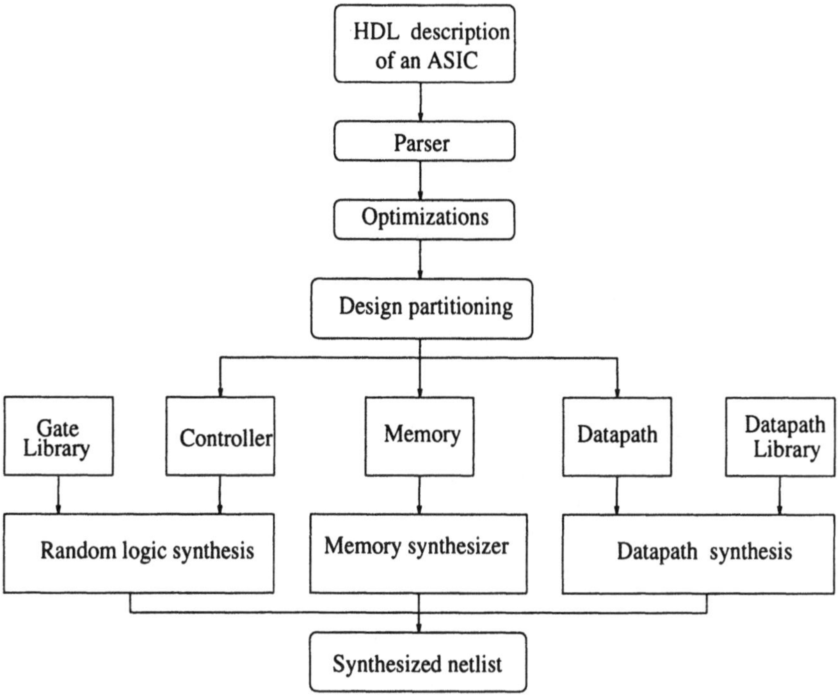


**Figure 2.8** *The ASIC synthesis using COMPASS' ASIC synthesizer.*

rapid and efficient convergence to the final IC design. Most recent high-level synthesis tools rank within this category (i.e., Hyper [40], Mistral 2 [43], Design Ware [53]).

RTL synthesis, on the other hand, provides more flexibility of design and less ease-of-use because designers have to deal with the design at the architectural and logic levels.

The full-custom design approach provides a great deal of flexibility. However, this comes at the expense of an unbearable degree of complexity that translates to a very long turnaround cycle.

Design methodologies that provide a comparable degree of flexibility with the custom design approach while still offering ease-of-use are based on efficient reusability. The ASIC synthesizer [54] proposed by COMPASS Design Automation is a datapath compiler that is based on use of basic arithmetic and bit-parallel operations. The design is subdivided into three blocks; datapath, controller, and memory. For efficiency of implementation, different tools are offered for the synthesis of each of these blocks. The datapath is described at the register transfer level (synthesis from the dataflow level is also possible) and is synthesized using the datapath synthesizer (or datapath compiler) tool. COMPASS' datapath synthesizer first parses the datapath inherent structure of the design into datapath blocks with the same width. Then, it maps every behavioral operation to cells from a library of datapath components such as adders, subtracters, and ALUs. Behavioral expressions that do not have corresponding cells in the datapath library are transformed through algebraic rules to expressions that have corresponding cells in the datapath library. Resource sharing is executed for mutually exclusive operations such as with the *case* statement. The synthesis process can also be guided using datapath synthesis directives that can control the partitioning of the datapath, the selection and sharing of resources, and other datapath compilation tasks (i.e., excluding an operation from being implemented in datapath). The controller is synthesized in random logic using the state machine synthesis option of the ASIC synthesizer. The logic implementation is optimized by recoding the state machine variable to one of four methods: (1) one hot state encoding (single-bit representation for each state), (2) logic adjacency, (3) state assignment by designer, or (4) random code. Memories are synthesized using the memory synthesizer provided by the ASIC synthesizer. Synthesis directives are offered to allow exploration and custom designing of different specifications and architectures. The flow diagram of COMPASS' ASIC Synthesizer is illustrated in Figure 2.8.

## *Core-based design*

Reuse that is considered at the functional level provides higher payoff in terms of ease-of-use because it takes the algorithm and the functional specficiation developer and the hardware/software designer much less work and time to cooperate effectively in the design, using off-the-shelf cores [55]). Such benefits have recently triggered an increased interest in reuse-based approaches [56, 57, 58, 59], especially with the rapid industrial acceptance of VHDL. Guidelines being formulated as part of the RASSP program are ensuring that DfR cores are interoperable and standardized to further this approach.

With the advances in VLSI technology, core-based systems can be integrated into a single chip or a MCM. The use of the dedicated core-based approach has proven to provide better processing efficiency than programmable DSP-based systems. This improved performance comes at the expense of a narrow range of applicability. As a remedy to this problem, architectures for connecting the

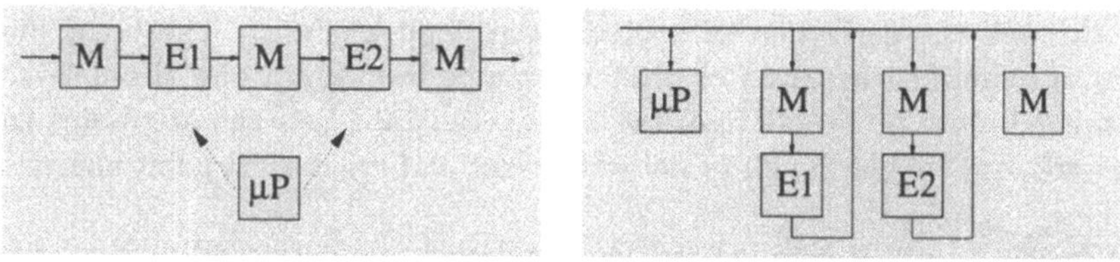


(a) (b)

**Figure 2.9** [60] *Core-based DSP architectures with microprocessors controlling the configuration and the data flow. M: Buffer Memory, E: Processing Engine, $\mu P$: Microprocessor. (a) The cores are connected based on the data flow of the application. (b) The architecture is built around a data bus.*

functional cores together, while providing flexibility have been independently proposed by Kasai and Minami in [60]. Such architectures require the use of a microprocessor that takes charge of data flow control and system configuration and initialization. They are broadly classified into two categories [60, 61, 62] (see Figure 2.9). In the first category, cores are connected based on the functional data flow of the application. The microprocessor can directly access each core and set the appropriate configuration and initialization procedures. The second category is based on a bus structure, where each core (including the host microprocessor) is connected to the system via a data/control bus. The second category provides additional capabilities to retarget the processing system to other applications, but at the expense of processing efficiency limitation. In the DSP applications considered in this book, the control aspects of the design were simple, but in more complex designs of reactive systems, some aspects of control and computation management (centralized or decentralized) require further experimentation.

## 2.4 MODEL FOR DESIGN COSTS

It is the general belief that FPGAs are considered cost-effective only when it comes to the prototyping for low-volume applications. ASICs, on the other hand, have been known to be cost-effective for high-volume production. In [46], Liu argued that these figures are based on a very simplistic cost model that only weighs the unit price and the non-recurrent engineering (NRE) cost. Such a model fails to capture many hidden costs of ASICs, such as; design time, production leadtimes, re-spin potential, risk, future cost reduction, and inventory. In order to illustrate the effect of these factors, Liu [46] developed a detailed cost model for a typical 10,000-gate design. The comparison is carried out through the derivation of a break-even point analysis. The break-even point is quantified in terms of product volume over which one alternative is considered more profitable than the other. Table 2.1 lists the attributes and costs adopted by Liu in [46] to derive the cost and break-even point analysis .

The model described here is based on the design attributes and product forecast listed by Table 2.1 [46]. This data is typical for a 10,000-gate design, according to trade reports from Dataquest and Integrated Circuit Engineering (ICE) provided by Liu.

## 2.4.1 Break-even analysis

Tables 2.2 and 2.3 list typical cost and time attributes of a 10,000-gate design, respectively. These attributes are given for both the ASIC and the FPGA alternatives, and are used to determine the true costs of ASICs and FPGAs.

### *Basic break-even analysis*

The basic break-even analysis formula is proposed in [46] as:

$$Basic\_break\_even = NRE + engineering + tools + (units * price). \quad (2.4)$$

This formula does not take into consideration the effect of the time-to-market. It is based only on the NRE costs, the engineering costs, and the unit price. The basic break-even equations for the FPGA and the ASIC approaches are illustrated by the following equations (using the data in Table 2.2 and Equation 2.4):

$$FPGA = \$25,000\ eng. + \$0\ NRE + \$10,000\ tools + (X\ units * \$39), \quad (2.5)$$

$$ASIC = \$79,000\ eng. + \$25,000\ NRE + \$10,000\ tools + (X\ units * \$13), \quad (2.6)$$

Solving for the break-even :

$$X_{ASIC} = 3038\ units,$$

This result implies that the ASIC approach is more beneficial than the FPGA approach only when the product volume is larger than 3038 *units*. This result,

**Table 2.1** Typical design attributes and product forecast for a 10,000-gate design.

| Design Attributes | | Product forecast | |
|---|---|---|---|
| 1. Number of gates in the design | 10,000 | 7. Year 1 | 5000 units |
| 2. Package type | PQFP | 8. Year 2 | 12,000 units |
| 3. Selling price of end system | \$2000 | 9. Year 3 | 5000 units |
| 4. Cost of end system | \$1100 | 10. Year 4 | |
| 5. Product life in months | 36 | 11. Year 5 | |
| 6. Engineering \$\$ per person per week | \$3000 | | |

however, does not take in consideration the time-to-market constraint. In order to take in consideration this important factor, the analysis should quantify the lost revenue due to a late product delivery.

## *Time-to-market analysis*

A delay in the delivery of a product results in losses in the revenue. These losses must be accounted for in the break-even analysis as a part of the total cost.

The percentage of lost revenue is a function of the delay in the delivery of the product and its life time. Figure 2.10 shows the triangle model, developed by Logic Automation (now owned by Synopsys), that illustrates the effect of late product delivery. The shaded area is proportional to the lost revenue. The percentage of lost revenues can be formulated as function of the product

**Table 2.2** Cost attributes for the ASIC and the FPGA approaches.

| Cost Attributes | ASIC | FPGA |
|---|---|---|
| 12. Engineering costs | \$79,000 | \$25,000 |
| 13. NRE | \$25,000 | \$0 |
| 14. tools | \$10,000 | \$10,000 |
| 15. Average price | \$13 | \$39 |

delivery delay $D$, and the product lifetime $2W$ as follows:

$$Percentage\_of\_lost\_revenue = [(D(3W - D)/2W^2)(100)]. \qquad (2.7)$$

Figures 2.11(a), and (b) illustrate the variation of the percentage of lost revenue as function of the product delivery delay, and the product lifetime, respectively. The shorter the product lifetime is, the larger the effect of the delay in the delivery becomes.

The total development time for the different design alternatives, as listed in Table 2.3, is the sum of items #17 through #26. The delay $D$ is calculated as the difference between delivering the product with an FPGA or an ASIC ($D = (32 - 11)/4 = 5.25\ months$). The percentage of lost revenue is relative to the net profit that is calculated as listed in [46]:

$$Net\ profit = Selling\ price\ of\ end\ system - (cost\ of\ system * volume/year). \qquad (2.8)$$

According to the data in Tables 2.2 and 2.3, and Equation 2.8, the net profit is:

$$Net\ profit = \$19,800,000.$$

The new break-even value is derived by adding the lost revenue to the total cost and re-solving Equation 2.4:

$$X_{ASIC} = 303,850\ units.$$

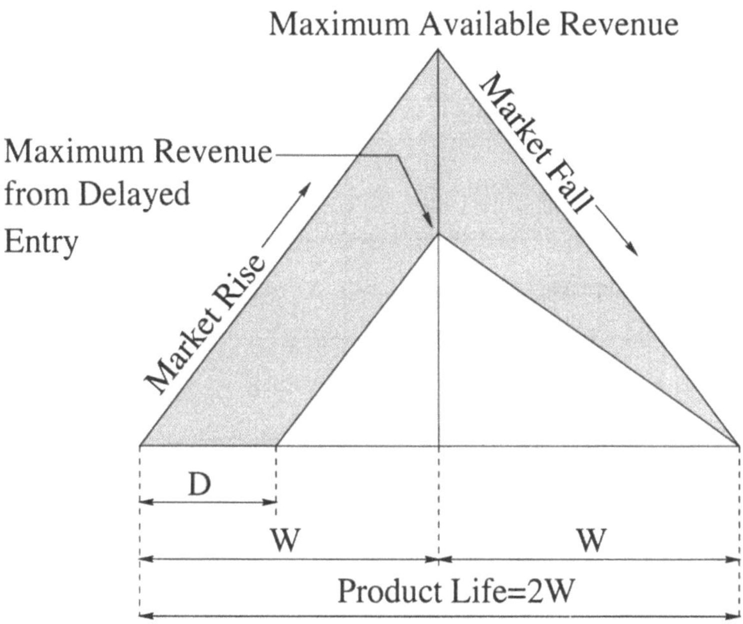


**Figure 2.10** *Synopsys' model illustrating the effect of late product delivery.*

This break-even point means that if the demand for the product is larger than 303,850 *units*, then the ASIC approach is better than the FPGA approach. The break-even point becomes even larger if other ASIC related costs (i.e., re-spin, and inventory) are accounted for. Surprisingly, the time-to-market factor accounts for about 90% of the total cost.

## 2.5 CONCLUSION

This chapter has reviewed the background required for this monograph. Design tools that allow a short design turnaround of ASICs are needed to compensate for a shrinking time-to-market and an increasing application complexity. Design abstraction techniques, efficient reusability, and automation appear to be the important techniques that can allow short design turnaround while conserving a cost/performance competitive advantage.

**Table 2.3** Time attributes for ASIC and FPGA approaches.

| Item | Time Attributes | ASIC (weeks) | FPGA (weeks) |
|---|---|---|---|
| 17 | Training | 2 | 1 |
| 18 | Capture | 3 | 2 |
| 19 | Simulation | 3 | 2 |
| 20 | Test-vector development | 6 | 0 |
| 21 | Place & route | 1 | 1 |
| 22 | Back annotation | 1 | 0 |
| 23 | Final simulation | 1 | 0 |
| 24 | Proto cycle | 2 | 0 |
| 25 | Qualification | 5 | 3 |
| 26 | Production leadtime | 9 | 2 |
| Total time | | 32 | 11 |

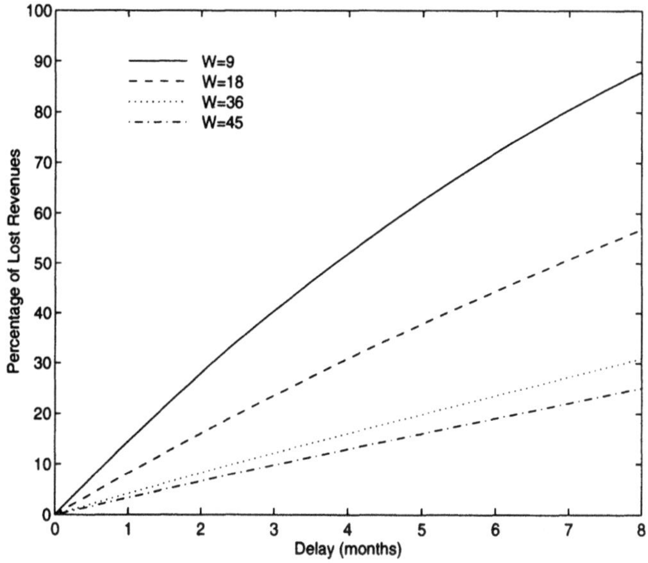


(a)

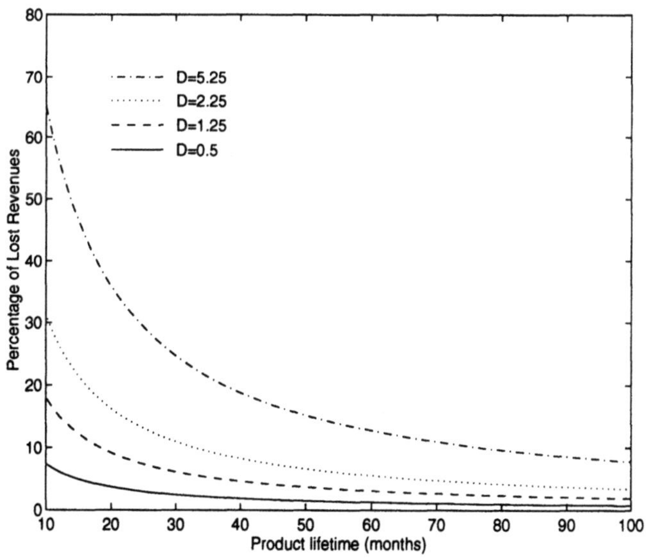


(b)

**Figure 2.11** *(a) Percentage of lost revenues as a result of a delay, D, in the product delivery. 2W is the product lifetime in months. (b) Percentage of lost revenues as a function of the product lifetime, for different product delivery delays.*

# 3

# VHDL-BASED DESIGN

Hardware design has recently undergone dramatic changes in design methodologies, especially with the proliferation of hardware description languages (HDLs) that promote the integration of the design methodology into a unified environment. Designs described in HDLs are kept at a more abstract level of representation than what traditional methods allow. HDL descriptions can take a variety of abstraction levels (See Figure 2.1). Synthesis is the step taken to translate the HDL description to a lower level of representation (i.e., the gate-level). Behavioral synthesis extracts an RTL (clock-level) structure from a behavioral HDL description. Practical approaches to behavioral synthesis set design restrictions (i.e., the use of pragmas in Mistral 2) in order to provide acceptable performance. While several HDLs exist, we recommend VHDL, which is also an IEEE standard (1076-1987/93).

This chapter starts with an introduction to the benefits of VHDL. Parameterized design with VHDL [15, 57] is then introduced through the design of a FIR filter. The constructs of the language that can support parameterized design are highlighted. The focus is on developing flexible and synthesizable VHDL cores in the Design for Reuse (DfR) phase. The built-in flexibility of parameterized design is at the essence of efficient reusability that can be implemented within the design methodology with VHDL. Systematic reusability is introduced at the end of this chapter. A more in-depth discussion on design with reuse (DwR) is presented in chapter 4.

## 3.1 VHDL MOTIVATION

Through the use of VHDL a number of benefits are accrued at various design stages: development, verification, synthesis, and testing. VHDL benefits include [27, 28, 30, 31, 32, 33, 29]:

- *Completeness* — Systems that are captured in VHDL can be maintained and upgraded for their lifetime period with no need to switch to another modeling language.
- *Technology Independence* — VHDL models can be kept technology independent until the very last stages of the synthesis process, providing the capability to retarget the same model to a variety of vendor technologies.
- *Portability* — Being a standard (IEEE Standard 1076-1987 [104]), VHDL allows the portability and the reuse of VHDL models across a wide range of design tools.
- *Abstraction levels* — VHDL supports a wide range of design abstraction levels and hierarchy. Hence, the same language is used for the entire design process, bridging the gap between design teams. Figure 3.1 illustrates an RTL and a behavioral descriptions for an RS flip flop and a Nand gate, respectively. The RTL model of the RS flip flop uses the Nand model as a component in its structure.
- *Department of Defense (DoD) compliance* — MILSTD 454 requires that ASICs designed for government projects be documented in VHDL.
- *Integration* — ASICs are integrated into the target system before that hardware implementations are attempted. The VHDL model of the ASIC is exercised in the system and design errors are caught at an early stage of the design process, while they are still easy to fix.
- *Excellent documentation* — Documentation is a very important issue with respect to future upgrades and maintenance. VHDL allows the documentation to be done during the design process (and not after), reflecting the true design intent.

The structured programming principles of the language provide for reliability, clarity, and excellent documentation. However, some new features such as those present in object-oriented languages may needed to handle the increasing complexity of large-scale VHDL implementations. The following section explores

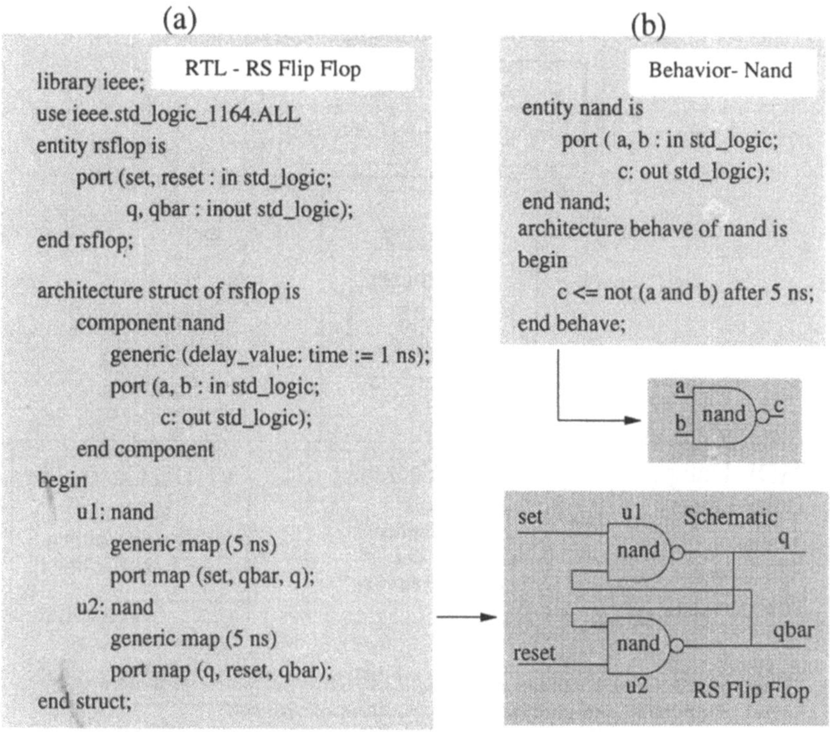


**Figure 3.1** *(a) RTL VHDL code for an RS flip flop. The component used by the RTL model is a Nand gate. (b) Behavioral VHDL model for a Nand gate.*

an parameterized component (for the DfR library) design with VHDL. The parameterization of designs introduces the flexibility needed to define an efficient reuse-based design methodology with VHDL.

## 3.2 PARAMETERIZED DESIGN WITH VHDL

The capabilities of the VHDL language to support parameterized design is the key to providing flexible cores that can adapt to different specifications. VHDL supports additional features that facilitate design encapsulation, inheritance,

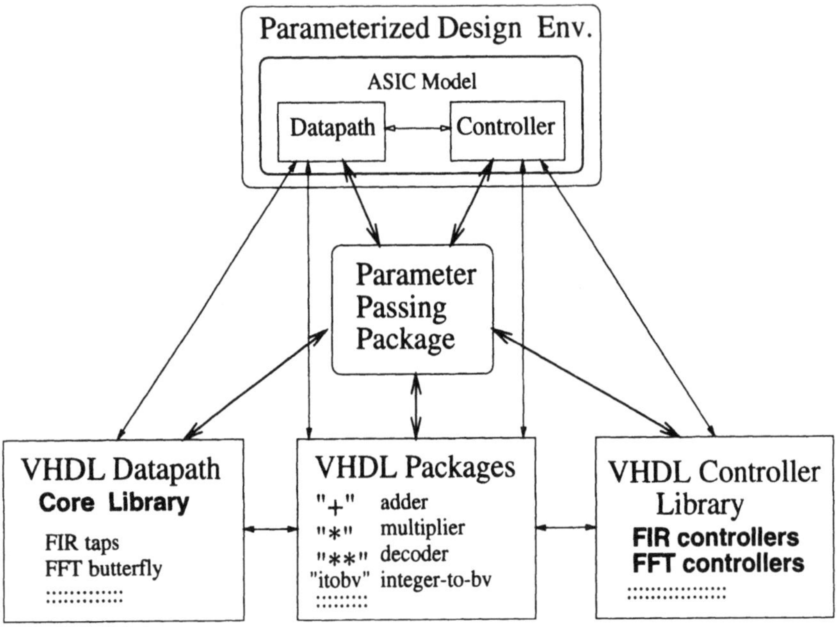


**Figure 3.2** *Parameterized design with VHDL— parameterized library of cores, macrocells, and packages supply reusable components.*

and reuse within the representation as will be outlined below. A description of VHDL is beyond the scope of this monograph.

VHDL packages are files that hold global data, functions, generics, and procedures to be shared among different designs. designs. Designs can reference the information stored in the package with little need to duplicate the information.

Generics are parameters that pass information to an entity. They can pass informations such as transition timing (type TIME) or size constraints (type INTEGER) [15]. For instance, in order to parameterize a VHDL core, designers can define integer generics that characterize the precision, the structure, or the pipelining depth of a hardware core. Other user-defined data types are also allowed.

Encapsulation reduces the number of details that a designer has to deal with, through the representation of the design as a set of interacting cores. The designer does not have to know how these cores work inside, but rather should focus on defining the appropriate interfaces between them. VHDL possesses powerful features for information hiding that translate into a great deal of system design simplification. Encapsulation is reinforced through the use of packages, functions, procedures, and entity declarations. An entity declaration is an interface to a VHDL core. It is similar to an icon in a graphical design environment.

Inheritance in VHDL is realized through parameter passing. The general structure of components and architectures is inherited by new designs, while restrictions are passed through a set of parameters. The parameters are passed to instantiate the design to fit the specifications of the target application. The procedures and functions defined within a package, also provide a great deal of abstraction and design reuse. Designers can just manipulate cores and macrocells instead of building designs from scratch.

Our approach to parameterized design with VHDL is illustrated by Figure 3.2. Parameterized VHDL libraries of cores, macrocells, and packages are the basis of the parameterized DwR design environment. An ASIC model is derived by reusing generic VHDL cores from a pre-designed VHDL library, or by using elementary parameterized datapath components from a defined VHDL package. In the following, the core-based design with VHDL is illustrated through the design example of an FIR filter.

Figure 3.3(a) illustrates the top of the hierarchy of a parameterized FIR design. The design is composed of a datapath and a controller. The parameters that introduce the built-in flexibility of this design are — *P, N, M, Q, Q_mux4*, which parameterize the data precision, the coefficient precision, the internal (accumulator) precision, the number of taps, and the multiplexing degree, respectively. The datapath is generated from a pre-designed VHDL FIR core, *filter4*, that is stored in the VHDL Datapath DfR Library illustrated in Figure 3.2. The controller is generated from a pre-designed VHDL FIR-controller core, *ctl_gen4*, stored in the VHDL Controller Library also illustrated.

The FIR datapath generator, *filter4*, is illustrated by Figure 3.3(b). The structure is based on the *generate* statement. The *generate* statement coupled with range parameterization, provides a powerful way to derive core generators in VHDL. The *generate* statement allows for structure replication in a telescopic way [15], familiar to most DSP hardware structures, where the same operation is repeated over a large stream of incoming data. Irregularities in the structure can also be handled through the use of conditional *generate* statements, as illustrated by the model of Figure 3.3(b), where the front-end tap of the FIR filter ($i = 0$) is different from the rest of the taps. Core generators using the *generate* statement result in correct-by-construction models that only require

```
--TOP LEVEL--

use work.passpara.all;
use work.fir_datatype.all;
use ieee.std_logic_1164.all;
library ieee;
library FIR;

entity top_level4 is
    port (datain : in bit_vector(P downto 0);
          clk, reset, work, load, switch: in bit;
          add: in integer range 0 to Q;
          coef : in bit_vector( N downto 0);
          acc : in bit_vector(M downto 0);
          dataout : out bit_vector(M downto 0));
end top_level4;
architecture gen_top_level4 of top_level
signal s1: bit;
signal s2: bit_vector(Q_mux4 downto 0);
signal s3, s4 : bit_vector(Q downto 0);

--CONTROLLER
component ctl_gen4
    port (clk, reset, work, load, switch: in bit;
          add : in integer range 0 to Q;
          w_bank : out bit;
          sel : out bit_vector(Q_mux4 downto 0);
          add1, add2 : out bit_vector(Q downto 0));
end component;

--DATAPATH
component filter4
port (datain: in bit_vector(P downto 0);
      clk, bank: in bit;
      sel : in bit_vector(Q_mux4 downto 0);
      add1, add2 : in bit_vector(Q downto 0);
      coef : in bit_vector(N downto 0);
      dataout: out bit_vector(M downto 0));
end component;
begin
    U1: ctl_gen4 port map (clk, reset, work, load,
                           switch, add, s1, s2, s3, s4);
    U2: filter4 port map (datain, clk, s1, s2, s3, s4,
                           coef,acc, dataout);
end gen_top_level4;
```

(a)

```
--DATAPATH--
entity filter4 is
    port ( datain: in bit_vector(P downto 0);
           clk, bank: in bit;
           sel: in bit_vector (Q_mux4 downto 0);
           add1, add2: in bit_vector(Q downto 0);
           coef: in bit_vector(N downto 0);
           acc: in bit_vector(M downto 0);
           dataout: out bit_vector(M downto 0));
end filter4;
architecture gen_filter4 of filter4 is
component cell9_front_end
port ( datain : in bit_vector(Pdownto 0);
       clk, bank: in bit;
       sel, regen_bank1, regen_bank2: in bit_vector
       (Q_mux4 downto 0);
       coef: in bit_vector(N downto 0);
       acc: in bit_vector(M downto 0);
       store_data: inout bit_vector(P downto 0);
       dataout : out bit_vector(M downto 0));
end component;
component cell9
port ( datain : in bit_vector(Pdownto 0);
       clk, bank: in bit;
       sel, regen_bank1, regen_bank2: in bit_vector
       (Q_mux4 downto 0);
       coef: in bit_vector(N downto 0);
       acc: in bit_vector(M downto 0);
       dataout : out bit_vector(M downto 0));
end component;
signal accz : regz;
signal s1: datax;
begin
accz(0) <=acc;
G: for i in 0 to Q_mux4 generate
    G1: if i=0 generate
        cell9_front_end0: cell9_front_end port map
        (datain,clk,bank,sel,add1(4*i+3 downto 4*i),
        add2(4*i+3 downto 4*i),coef,accz(i),s1,accz(i+1));
    end generate;
    G2: if (i/=0) generate
        cell9i: cell9 port map (s1, clk, bank, sel,
        add1(4*i+3 downto 4*i),add2(4*i+3 downto 4*i),
        coef,accz(i),accz(i+1));
    end generate;
    dataout <= accz(Q_mux4 +1);
end generate;
end gen_filter4;
```

(b)

**Figure 3.3** *Parameterized family of VHDL cores of a FIR filter.*

```
package DATATYPE4 is
type state is (S0, S1, S2, S3a, S3b, S3c,
            S3d,S4a, S4b, S4c, S4d, S5);
end DATAYPE4;
use work.passpara.all;
use work.DATATYPE4.all;
entity ctl_gen4 is
    port (clk, reset, work, load, switch: in bit;
        add: in integer range 0 to Q;
        w_bank: out bit;
        sel: out bit_vector(Q_mux4 downto 0);
        add1, add2: out bit_vector(Q downto 0));
end clk_gen4;
architecture behavior of ctl_gen4 is
signal CURRENTSTATE: state;
signal NEXTSTATE: state;
begin
ctl_gen4_process: process(CURRENTSTATE, work,
                          load, switch, add, sel)
begin
    case CURRENTSTATE is
        when S0 =>
            if load = '0' then
                add1 <= itobv(0,Q+1);
                add2 <= itobv(0,Q+1);
                sel <= "0000";
                if switch = '0' then
                    w_bank <= '0';
                else
                    w_bank <= '1';
                end if;
                NEXTSTATE <= S0;
            else
                if switch = '0' then
                    add1 <= itobv(2**add, Q+1);
                    add2 <= itobv(0, Q+1);
                    w_bank <= '0';
                    sel <= "0000";
                else
                    add1 <= itobv(0, Q+1);
                    add2 <= itobv(2**add, Q+1);
                    w_bank <= '1';
                    sel <= "0000";
                end if;
                NEXTSTATE <= S1;
            end if;
        when S1 => ::::::::::::::::::::::::::::::
        when S2 => ::::::::::::::::::::::::::::::
        :::::::::::::::::::::::::::::::::::::::::
        when S5 => ::::::::::::::::::::::::::::::
        end case;
        end process ctl_gen4_process;
        ctl_gen_seq: process(clk,reset)
        begin
            if(To_X01(reset='0') then
        CURRENTSTATE <=NEXTSTATE;
            end if;
        end process ctl_gen_seq;
        end behavior;
```

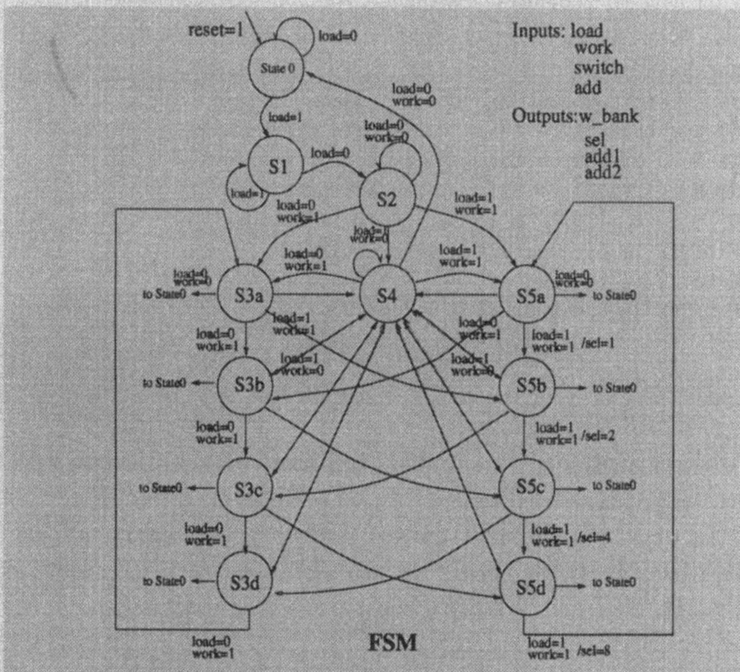


**Figure 3.4** *Parameterized VHDL core for a FIR controller.*

to be checked for typical cases and boundary conditions. This is because the structure of such core generators has already been validated for some specific instantiation cases, and because they re-use pre-designed and pre-verified cores. The module generator of Figure 3.3(b) uses the cores — *cell9_front_end*, and *cell9*. These are "multiplexed-by-a-factor" of four FIR taps. The VHDL model controller for this filter is illustrated by Figure 3.4. The controller is a 12-state finite state machine (FSM) . The states of the FSM are declared using abstract modeling through the enumerated type [64]. The FSM is composed of two processes, a concurrent process (*ctl_gen4_process*), and a sequential process (*ctl_gen_seq*) as is well known to VHDL users. The FSM model ripples through the defined states based on the state of the *CURRENTSTATE* signal. The *case* statement is used to make the appropriate procedure execution selection. The transitions between the different states are defined in the concurrent process. The sequential process is sensitive to the clock and the reset signals. It assigns the *NEXTSTATE* signal to the *CURRENTSTATE* signal on the rising edge of the clock.

The VHDL package that passes the design parameters to the different hierarchy levels of the design might look like this:

```
PACKAGE passpara is
CONSTANT Q : INTEGER := 63;       -- number of taps - 1
CONSTANT P : INTEGER := 12;       -- data precision
CONSTANT N : INTEGER := 16;       -- coef precision
CONSTANT M : INTEGER := 32;       -- acc precision
CONSTANT Q_mux4 : INTEGER := 16; -- (Q+1)/4
SUBTYPE DATAX is bit_vector( P downto 0);
SUBTYPE DATAY is bit_vector( P + N + 1 downto 0);
SUBTYPE DATAZ is bit_vector( M downto 0);
TYPE REGZ is array(Q + 1 downto 0) of DATAZ;
END passpara;
```

This package instantiates a "multiplexed-by-four" 64-tap FIR filter with 12-bit data precision, 16-bit coefficient precision, and 32-bit internal precision. By fixing the pre-defined parameters the architecture and precision of the design are instantiated to fit the application specifications. The level of hierarchy in this design is illustrated in Figure 3.5.

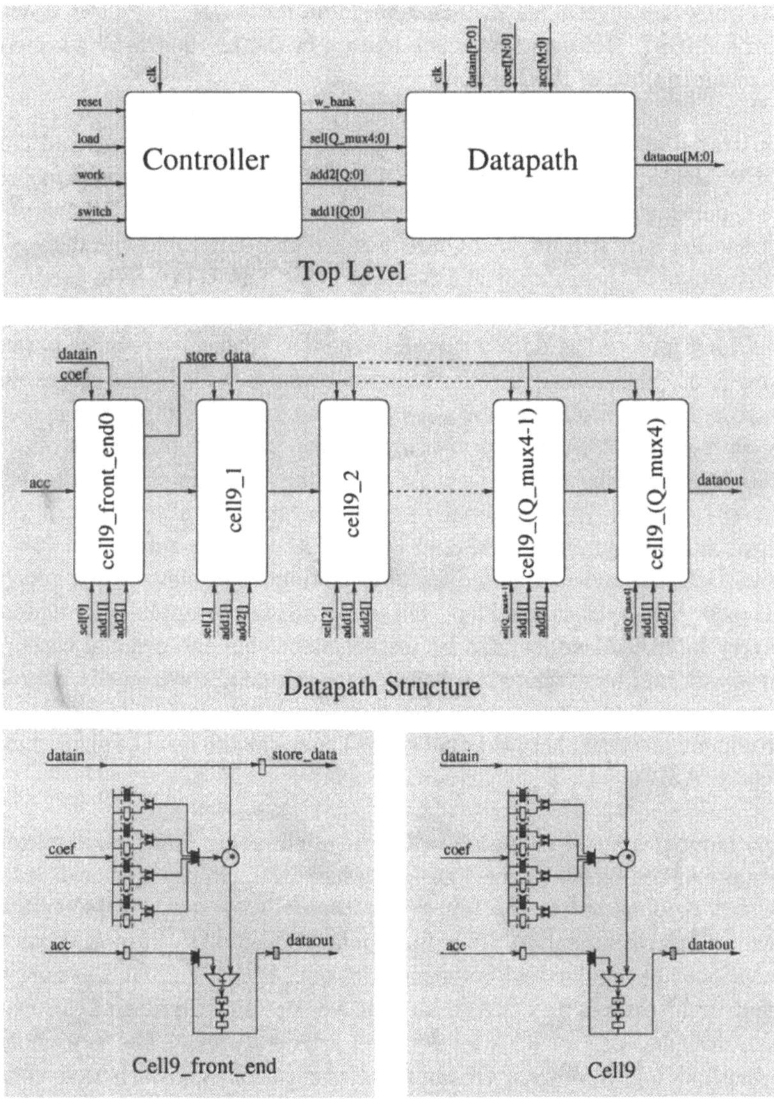


**Figure 3.5** *Illustration of the design levels of a parameterized FIR filter core. At the top level the design is composed of a controller and a datapath. The datapath structure is derived from a parameterized module generator (i.e., the number of taps is a parameter). The functional cores (FIR taps) of the datapath are also illustrated. Parameterization is introduced at all design levels.*

## 3.3 SYSTEMATIC DWR

VHDL possesses several features that promote reuse (i.e., packages, functions, and procedures). However, reuse is a feature of the methodology adopted and not a responsibility of the language.

As illustrated above, VHDL possesses the required characteristics and flexibility (abstraction, encapsulation, generalization, inheritance, reuse) to provide efficient parameterized and reusable core generators for the DfR library. Nevertheless, there are still no clear guidelines and standards that can take systematic advantage of these important characteristics, and help VHDL programmer organize complex programs more efficiently. For example, how to link the processing functions to the data structure ?, how to realize systematic parameter passing ?, or, how to organize for efficient reusability ?. These issues narrow down to a fundamental question — what is the appropriate synthesis methodology with VHDL ?. Some object-oriented proposals aim to put the burden on the language itself by extending it to adequately support object-oriented constructs [67, 68, 69]. This approach requires an additional effort from vendors in terms of programming style and conceptual understanding. In [68] Powley and DeGroat have proposed an object-oriented framework for generating VHDL code for the class of adders. Objects, such as the ripple-carry adders and the carry lookahead adder, can be instantiated from the original class. Such a framework can be extended to handle more classes of arithmetic operations. The products of such framework can also be used as leaf cells for the design environment proposed here that considers classes at the level of algorithms and objects as ASICs.

In this monograph, we define a DwR/DfR methodology that takes systematic advantage of the parameterized design with VHDL. Reuse is considered at all levels but emphasized at the functional level. The result is a systematic approach and environment to designing domain-specific DSP applications. DSP applications are regular and deterministic (e.g., FIR, FFT, DCT, etc.). Hence efficient architectures and design alternatives are either known beforehand or can be designed at the RTL level (novel architectures). The systematic design methodology follows a tree-based classification approach that considers one domain-specific application at a time. Once an algorithm is chosen (e.g., the FIR algorithm), design alternatives are listed horizontally, and a variation of implementations for each alternative are listed vertically. Design alternatives represent a variation of architectures (sequential, parallel). Every design alternative spans vertically to a variety of implementation issues (e.g., pipelining, multiplexing). Reuse is organized such that at the top of the tree a design architecture calls pre-designed, pre-verified domain-specific macrocells and cores (e.g., an FFT architecture calls butterfly datapath cores). These cores themselves call upon datapath cores of smaller sizes. The parameters that instantiate the design spread vertically into the tree structure. Chapter 5 describes results from a number of implementations of the DwR methodology.

# 4

# DESIGN FOR REUSE

This chapter starts with an introduction to the core-based DfR/DwR methodology proposed in this monograph. The issues involved in defining an efficient reuse-based design environment are investigated and defined. Later sections describe the different *design views* that constitute the proposed design methodology. A cost model that compares the relative costs for proposed and existing approaches to ASIC design is also presented.

## 4.1 INTRODUCTION

This chapter proposes a novel design methodology for application-specific integrated circuits (ASICs) based on *systematic* functional core reuse [70]. Efficient functional reusability is derived from the built-in flexibility of parameterized design with VHDL [15], and its many flexible constructs and characteristics (i.e., abstraction, encapsulation, inheritance, and reuse through *types*, *attributes*, *packages*, *procedures*, and *functions*). Efficient and rapid design exploration is realized through a limited but realistic design space exploration bounded by the proposed reusable datapath library of functional cores. In short form, we call this design methodology, Functional-core ASIC design (FASIC) . Component identification, classification, storage, and retrieval, within the proposed environment, are handled systematically. A Tcl/Tk Graphical User Interface (GUI) is offered for this purpose. The designer is able to use one of two design lines — (a) the Automatic Design Line (ADL) is a collection of turn-key tools for deterministic DSP algorithms (i.e., FFT, FIR, Viterbi). Each provides wide design exploration capabilities through a tree-based search algorithm of a variety of reusable architectures and cores, (b) The Functional Design Line (FDL) is an environment where applications are designed at the register-transfer level (RTL) using functional decomposition onto the core library. defined functions are then mapped to the corresponding functional cores from the reusable DfR library.

A detailed cost/time model is also proposed, and a comparison between FA-SICs (products of our design approach), ASICs, and field programmable gate arrays (FPGAs) is conducted based on the derivation of a break-even analysis. The break-even point is quantified in terms of product volume over which one alternative is considered more profitable than the other [46].

A DfR library is more than a set of pre-designed cores. The DfR library developer and designer are faced with a number of questions that need to be resolved:

- *How does one use these cores within the design environment ?* — This question relates to of ease-of-use and involves the identification and understanding of the functional components that can be used to implement the target application's functionality and performance.
- *How flexible are these cores ?* — When appropriate cores are identified, there is always a need to customize them to fit certain design specifications (i.e., data format, precision). For the sake of synthesis efficiency and design simplicity it is important to determine what parameters should be made accessible to the user and what should be hidden.
- *What performance can one expect from the implementation ?* — It is crucial that the designer be able to estimate the performance of the design (area, speed, power) at the early stages of the design process. Hence, accurate information about the performance of the reusable cores must be provided. Also, a minimum amount of design exploration is required to investigate better design alternatives. This requires that the DfR library provide a number of cores that have the same functionality but different performance and form measures.
- *How about design upgrades, portability and configuration management ?* — With flexible specifications, designers want the freedom to make rapid design changes or use different tools. They do not want to get stuck with a design that cannot be ported to various technologies and simulation & synthesis tools. VHDL, as a chosen standard for hardware description languages, removes some of the risk associated with this flexibility.

- *Can one add additional cores to the DfR library ?* — Designers should be able to contribute to the enlargement of the DfR core library. However, contributions must conform to some specified guidelines such that the same ease-of-use and flexibility standards are enforced.

The reuse-based design environment proposed in this chapter takes into consideration these questions, and broadly classifies theses issues into three design views:

1. The *core or module* design view is concerned with the design of a reusable datapath library of cores for DSP algorithms. This is the **design-for-reuse** (DfR) cycle (off-design-cycle) that involves all the considerations that have to be taken in order to supply the environment with a reusable library that provides ease-of-use and efficiency. The emphasis here is on identifying typical cores and increasing their reuse potential through abstraction and parameterization.
2. The *system* design view defines the structure of the design environment and the approach to specifying DSP systems using the reusable pre-designed core library as a foundation. This is the **design-with-reuse** (DwR) cycle (in-design-cycle) that involves the selection of the appropriate reusable cores from the library and the tailoring of these cores to meet the desired specifications.
3. The *simulation* design view relates to the verification of the functionality and the conformity of synthesized performance to the design specifications and customer requirements.

## 4.2 CORE DESIGN VIEW

This design view corresponds to the Design-for-Reuse (DfR) cycle. The major concerns of DfR phase are:

1. Identifying reusable cores for DSP applications. This task corresponds to defining the flexible units of behavior for DSP applications. These flexible units of behavior of sufficiently high granularity are called cores. For instance, a typical core for a Viterbi decoder application would be the the Add-Compare-Select Unit (ACSU) [71].

2. Increasing the potential of reuse of the identified cores in order to encompass a wide design exploration space.
3. Assuring a certain degree of ease-of-use of the proposed reusable cores.

The last two concerns conflict with each other, because increasing the potential of reuse of DfR cores implies that many features have to be implemented within a single core. On the down side, this increases the complexity of designed cores, limits their wide usage.

Given these concerns, a key issue in the module design view is how should the library cores be *scaled*? — vertically (size wise), horizontally (feature wise), or both. Size scaling accommodates Design-with-Reuse (DwR) because designers will be manipulating only a few number of cores instead of a larger number of components of smaller sizes. However, as cores grow in size, the potential of reuse decreases because the larger the size of a dedicated core, the more narrow its domain of application becomes [65, 66]. Programmable cores based on kernels of DSP chips are an attractive option that maintain the flexibility of programmability.

### 4.2.1 Core identification

A compromise between horizontal and vertical scaling is adopted. Cores that populate the datapath DfR library are identified as being algorithm-specific, meaning that the minimum component size characterizes the component to be **application-specific** as opposed to being general purpose. That is, the size of a component is increased until it becomes algorithm-specific. For example, a single multiplier is too small to be considered as an application-specific core. However, a butterfly core can target FFT applications. The butterfly operation is, thus, an elementary domain-specific core that builds up an FFT application.

Once the size of the core is chosen, other versions of the same core are derived. The goal is to encompass a wide variation of design features and architectural styles of the target DSP algorithm. For example, the elementary core for designing FIR ASICs is one FIR tap. Several architectural styles are then considered for building this FIR tap core. The sequential structure simply corresponds to a multiply-accumulate block, while the direct-form structure is targeted for building parallel FIR filters. The transpose-form structure is desirable for a parallel implementation because it has a short critical path. For each architectural style considered, several versions of the target core are provided in synthesizable VHDL. These design versions encompass a variety of design techniques at the architectural and the micro-architectural levels (pipelining, multiplexing) in order to accommodate a wide range of design specifications.

Within this context, a reusable core is defined as a domain-specific building block that can embrace a variety of structures and features. The model that can represent reusable cores is a triplet:

*(Domain, Structure, Feature).*

The domain $D$ defines a target DSP algorithm for the reusable core (i.e., FIR, FFT, DCT, Viterbi). The structure $S$ defines the architecture embraced by the reusable core. For instance, a direct-form structure or a transpose-form structure, for an FIR application. In the case of an FFT application, the structure $S$ might instantiate a decimation-in-time or a decimation-in-frequency structure. The feature $F$ states the variations of the reusable core for fixed domain and structure (i.e., degree of multiplexing, level of pipelining). This abstract representation of hardware components is a high level reference to actual hardware components in the reusable DfR library. Designs can be retargeted to fit different specifications with no need to change the source code. The parameterized design with VHDL allows this flexibility of representation of components.

## 4.2.2 Core flexibility

The question that is raised here is — *what parameters should be made accessible to the designer to customize reusable cores in order to fit the specifications of his or her target application ?*. Several studies have shown that what designers are most concerned about are (1) ease-of-use of the reusable components (understanding), and (2) the performance that results from using these reusable cores. What they are least concerned about are (1) How the reusable cores are written, and (2) the potential of reuse of these cores (i.e., if a designer selects a reusable block for a DCT application, she would not be too concerned whether the same core can also be used for a Viterbi algorithm).

Design parameters of a core can instantiate:

- word precision,

- functionality,
- data representation (i.e., 2's complement, unsigned),
- structure (architectural issues),
- features (variations of pipelining and multiplexing),
- others.

For reasons of ease-of-use and performance (efficiency of synthesis), only the precision of the different datapath cores, and the depth of an architecture style (i.e., order of a filter, or a size of an FFT application) are parameterized within one VHDL core. This does not prevent the designer from selecting different structures or features. The functionality, structure, and features are implemented in different datapath cores (or VHDL models). That is, a different functionality (feature) is implemented in a separate datapath core. For example, a "multiplexed-by-two" FIR tap, and a "multiplexed-by-four" FIR tap are implemented in two different datapath cores. Feature variation of the proposed cores attempts to capture as much as possible of design exploration in order to fit a wide range of data rates, while at the same time optimizing area and power. Issues such as core identification, classification, storage, and retrieval are handled systematically by the proposed design environment. Every datapath core possesses a set of parameters that characterizes the precision of the different datapath sub-blocks within the datapath core. The VHDL code illustrated by Figure 4.1 represents the entity of a front-end FIR tap, where $P$, $N$, and $M$ are parameters for customizing the precision of the different datapath cores. The design parameters are passed via a VHDL package to instantiate the cores to accommodate design specifications.

### 4.2.3 Portability and other VHDL benefits

The datapath library proposed here is designed in portable, simulatable, synthesizable VHDL code. This allows for exercising ASICs in board-level environments before hardware prototypes are attempted such that design flaws are uncovered early during the design process. VHDL representation also allows for excellent documentation capabilities, resulting in easy upgrading and maintaining, as mentioned earlier.

```
entity cell1_front_end is
-- P, N, and M are respectively data, coefficient,
-- and accumulator precision.
port ( datain    : in bit_vector (P downto 0);
       clk, bank, en_REG1, en_REG2, en_output : in bit;
       coef : in bit_vector ( N downto 0 );
       acc :  in bit_vector ( M downto 0 );
       store_data : inout bit_vector (P downto 0 );
       dataout : ou bit_vector ( M downto 0);
     );
end cell1_front_end;
```

**Figure 4.1** *VHDL entity of an FIR front-end core.*

VHDL also provides the ability to describe circuits and systems at different levels of abstraction. The VHDL code illustrated by Figure 4.2 represents a (dataflow abstraction level) transpose-form structure front-end FIR tap.

## 4.2.4 Organization of the DfR library

The library environment proposed is based on a hierarchical structure of several levels (see Figure 4.5). At the lowest hierarchy level (**LEVEL-0**), elementary *datapath-components* such as adders, subtracters, multipliers, registers, register files, shift registers are provided. The next hierarchy level (**LEVEL-1**) uses these elementary datapath components to build-up domain-specific *datapath cores* such as LMS blocks, FIR components, and FFT butterfly cells. At a higher hierarchy level (**LEVEL-2**), VHDL *architecture files* are provided. Such files specify how these datapath cores must be configured and connected together. Architecture files can provide both classic and novel architectures. For example, Figure 4.3 illustrates a VHDL architecture file for a transpose-form FIR filter. The final structure of the synthesized filter is similar the parallel FIR architecture of Figure 4.4(a), where the library components used are *cell1_front_end* (front-end FIR-tap), and *cell1* (FIR-tap). The entity and the

```
architecture behave of cell1_front_end is
signal REG1 : bit_vector ( N downto 0 );
signal REG2 : bit_vector ( N downto 0 );
begin
        process ( clk )
        variable mul : bit_vector ( (P+N+1) downto 0 );
        variable tmp1 : bit_vector ( N downto 0);
        variable mul1 : bit_vector ( M downto 0);
        variable tmp : bit_vector ( N downto 0);
        begin
            if clk'event and clk = '1' then
                if en_output = '1' then
                store_data <= datain;
                end if;
                if en_REG1 = '1' then
                REG1 <= coef;
                end if;
                if en_REG2 = '1' the
                REG2 <= coef;
                end if;
                tmp := acc;
                if bank = '1' then
                tmp1 := REG2;
                end if;
                mul := store_data * tmp1;
                mul1 := sign_entend ( mul, M+1 );
                if en_output = '1' then
                dataout <= mul1 + tmp;
                end if;
            end if;
        end process;
end behave;
```

**Figure 4.2** *VHDL description of a transpose-form FIR tap.*

```
entity filter is
architecture gen_filter of filter is
component cell1_front_end
     port ( datain : in bit_vector (P downto 0);
            clk, bank, en_REG1, en_REG2, en_output : in bit;
            coef : in bit_vector ( P downto 0 );
            acc : in in bit_vector ( M downto 0 );
            store_data : inout bit_vector ( P downto 0 );
            dataout : out bit_vector ( M downto 0 ) );
end component
component cell1
     port ( datain : in bit_vector (P downto 0);
            clk, bank, en_REG1, en_REG2, en_output : in bit;
            coef : in bit_vector ( P downto 0 );
            acc : in in bit_vector ( M downto 0 );
            dataout : out bit_vector ( M downto 0 ) );
end component
signal accz : REGZ;
signal s1 : DATAX;
begin
accz(0) <= acc;
G: for i in 0 to Q generate
     G1: if i = 0 generate
            cell1_front_end0 : cell1_front_end port map
            (datain, clk, w_bank, add1(i), add2(i), en_output,
            coef, accz(i), s1, accz(i+1));
      end generate;
      G2: if (i/=0) generate
            cell1i : cell1 port map (s1, clk, w_bank, add1(i),
            add2(i), en_output, coef, accz(i), accz(i+1));
      end generate;
     dataout <= accz(Q+1);
end generate;
end gen_fitlter;
```

**Figure 4.3** *VHDL architecture file for a transpose-form FIR filter. The Generate statement is used to connect the specified cells together.*

architecture for the core *cell1_front_end* are illustrated by Figures 4.1 and 4.2, respectively.

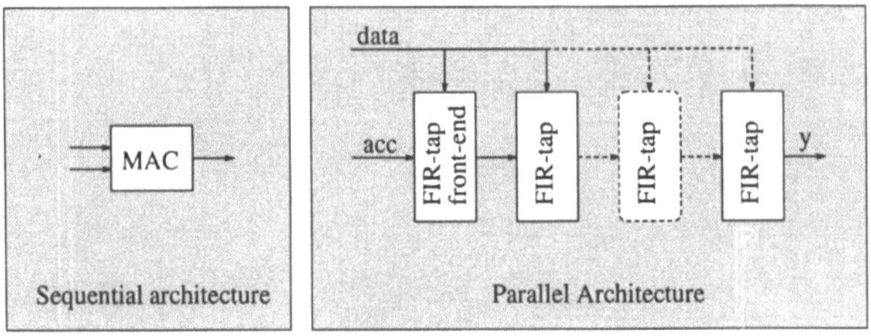


(a)

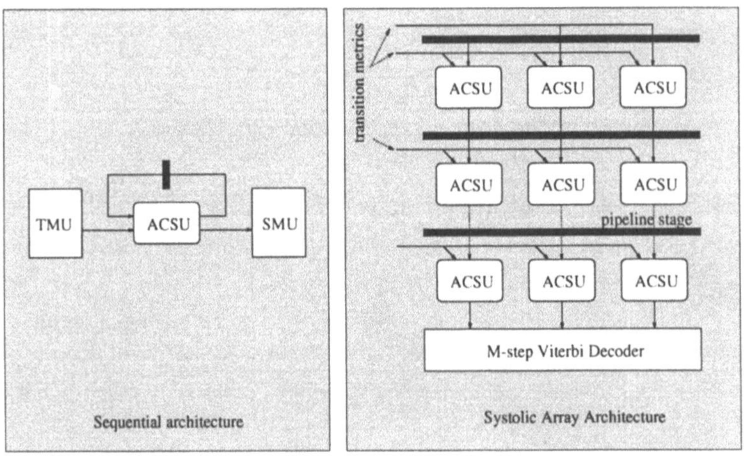


(b)

**Figure 4.4** *Sequential and parallel architecture files: (a) FIR algorithm. (b) Viterbi algorithm.*

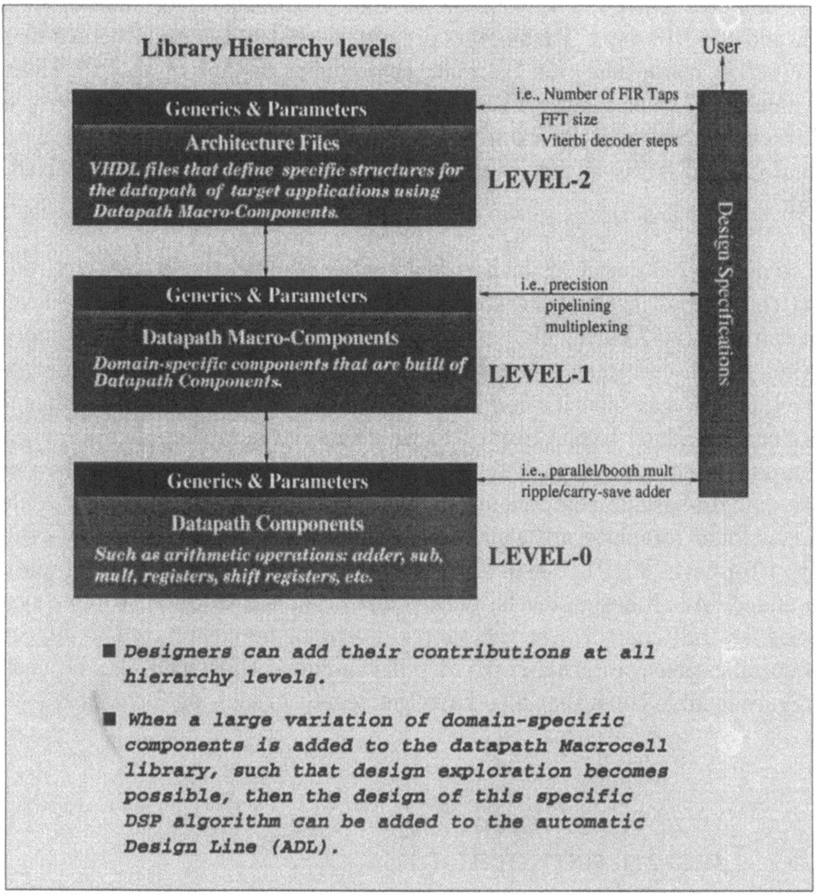


**Figure 4.5** *Organization of the Proposed Library.*

The described filter is built-up of two datapath macro-cores: *cell1_front_end*, and *cell1*. These cells are used in numbers to meet the number-of-tap specification. The structure of the filter is defined using a *generate* statement that connects the cells together.

Figures 4.4(a), and (b) [72] illustrate examples of sequential and parallel architecture files for the FIR and the Viterbi algorithms. These files use components from the core library such as the add-compare-select unit (ACSU), the multiplier accumulator (MAC), transition metric unit (TMU), survivor memory unit (SMU), and the FIR-taps. Parameters are also passed to the architecture files of the cores to instantiate, for example, the number of taps of an FIR filter, or the number of steps for the Viterbi decoder. At the top of the hierarchy, a VHDL program connects the datapath and the controller of the target design. The final desired ASIC is the result of the synthesis of this top-level VHDL program.

The core designer can add his or her own contribution to the library environment at the various levels of hierarchy. At the elementary datapath core level he/she can add (customize) higher performance elementary datapath components (i.e., Booth multiplier, carry-look-ahead adder). At the datapath macro-core level, new domain specific components that use elementary datapath components can be added to the library to widen its domain of application (i.e., add-compare-select unit for a Viterbi decoder). In order to facilitate the use of these domain-specific components for system designers, an architecture file and a controller template are also added at the architecture hierarchy level. Finally, a top-level VHDL file that connects the controller and the datapath is also added. At all design levels, parameterization is introduced to give system designers full control over the target design to be customized to fit the design specifications. Internal cores of programmable DSPs can also be used to configure flexible cores that also have the benefit of high performance.

### 4.2.5 Design exploration

Every core in the proposed DfR library is provided with full information about the expected area, speed, and power for different technologies and architectural styles. Estimation of design cost information about applications that use cores from the library are easily drawn. Figure 4.6 illustrates design cost information of increasing order digital filters designed from the cores provided by the datapath DfR library.

The variation of the library cores is organized such that they cover a wide range of design specifications. For example, a variation of the degree of time multiplexing is targeted to optimize the area, while increasing the level of pipelining (inter, and intra-primitive cells) is meant to adapt the datapath cores for high data rates. Power dissipation is customized by reducing the supply voltage (i.e., the throughput is kept constant by means of combining pipelining with a greater circuit parallelism), tagging the design to a low power primitive cell li-

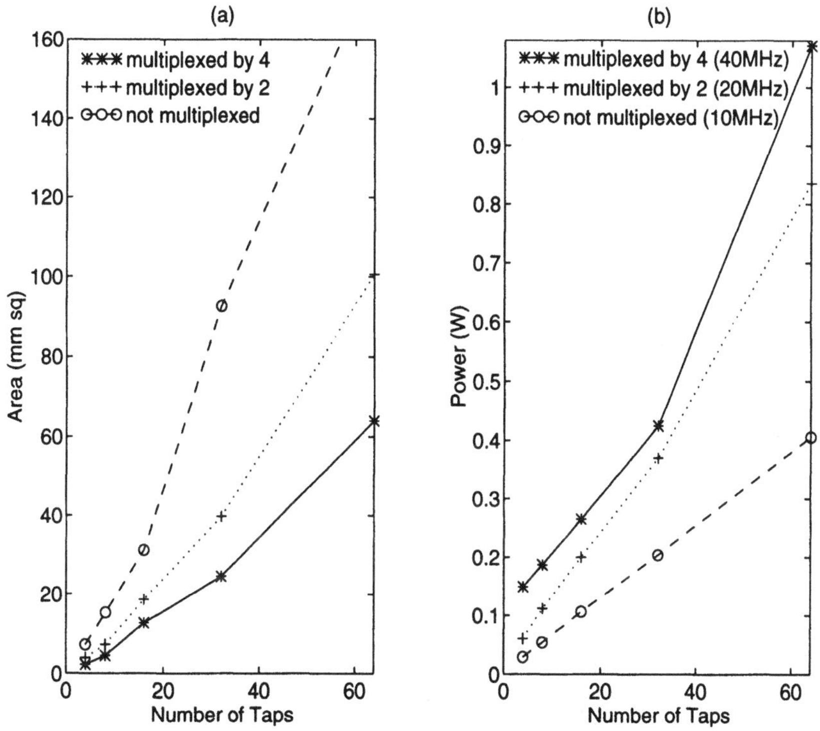


**Figure 4.6** *Area and Power estimations in* 0.8$\mu$ *CMOS technology for FIR filters of different degrees of multiplexing, one degree of pipelining and a data rate of* 10*Mbaud. (a) Area estimation as a function of the number of taps (b) Power estimation as a function of the number of taps.*

brary, or choosing from a provided variety of multiple parallel functional units. Recently, efforts at the DSP Laboratory at Georgia Tech have implemented a *fuzzy logic* front-end to assist with the design exploration process with encouraging results.

## 4.3 SYSTEM DESIGN VIEW

This design view corresponds to the Design-with-Reuse (DwR) cycle. It is mainly concerned with the use of the proposed reusable datapath library and environment in order to customize DSP-ASICs. The goal is a reduction in the design turnaround cycles of ASICs by at least a factor of "4X". At the same time, the system design view provides ease-of-use, high application flexibility and integrability into larger systems. The designer is invited to use one of two design lines:

- The *automatic design line* (ADL) is a collection of turn-key tools for deterministic DSP algorithms (i.e., FFT, FIR, Viterbi). Each provides wide design exploration capabilities through a tree-based search algorithm of a variety of reusable architectures and cores. Core identification, classification, storage, and retrieval are handled systematically. A Tcl/Tk [73] graphical user interface is offered for this purpose.

  Unlike conventional turn-key approaches which provide for ease-of-use but lack flexibility, the proposed ADL tool favors both features. The flexibility is introduced through the parameterized VHDL cores and within the design methodology that allows for a design space exploration bounded by the proposed datapath library. Ease-of-use is introduced through automation. The flow diagram of the ADL tool is illustrated by Figure 4.7.

- The *functional design line* (FDL) is an environment where applications are designed at the register-transfer level (RTL) using functional decomposition. The defined functions are then mapped to the corresponding functional cores from the DfR library.

### 4.3.1 Systematic reuse

Systematic reuse is an essential factor for a wider acceptance of reuse-based design. Systematic reuse is implemented through automatic selection of the appropriate building blocks or cores, based on the front-end specifications of the system designer and customer. Systematic reuse at the architectural level is implemented through a tree-based search algorithm that identifies the desired cores based on the high-level representation, *(Domain, Structure, Feature)*, and the original specifications. A design path is chosen given a set of design parameters. Figure A.1 in the appendix illustrates the design automation for

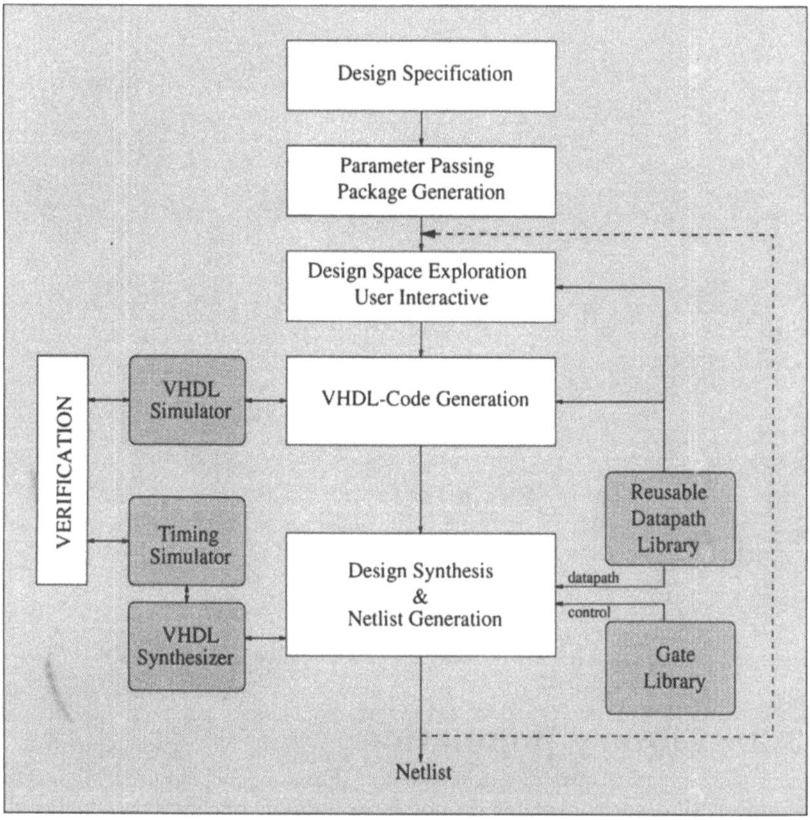


**Figure 4.7** *Flow diagram of the ADL tool.*

the FIR algorithm (see Figure 4.9 for the design of FFT applications). At every design step several design alternatives are proposed, but only one path is selected. The chosen path is the one that fits the design specifications the best. The leaves of the design tree are the elementary datapath cores that are used by the domain-specific components of the macro-cell datapath library. The DfR library is populated with a wide range of domain-specific cores facilitating efficient design space exploration.

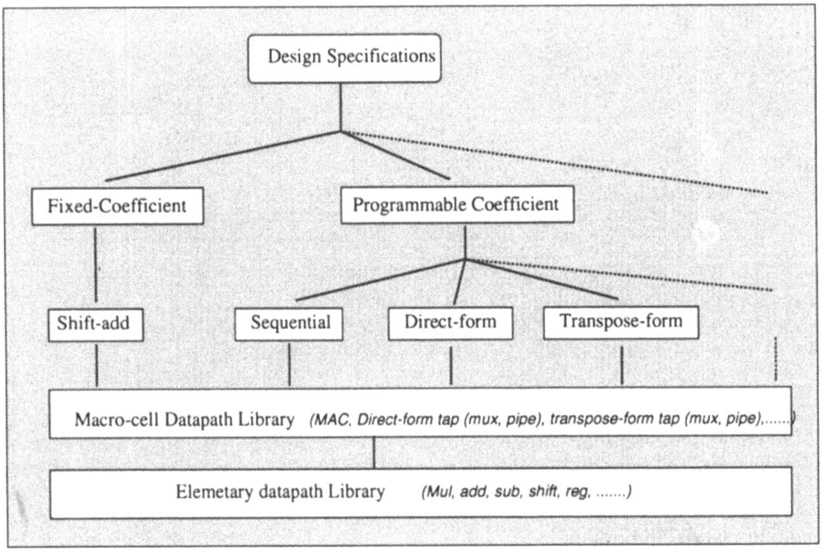


**Figure 4.8** *A tree-based design exploration of FIR filters.*

## 4.3.2 Library population

For some DSP algorithms that do not have domain-specific cores in the datapath library, designers can still use the elementary datapath library elements (such as adders and multipliers) to build new domain-specific datapath cores. These cores can then be added to the macro-core datapath library. Once a range of these domain-specific cores is available in the DfR library design exploration becomes possible and efficient. Additions to the architectural design exploration for the supported algorithm are also possible. Designers will just have to add architecture files to the environment. The same macro-cell core can be used by different architecture files in order to widen the design exploration space (see Figure 4.4).

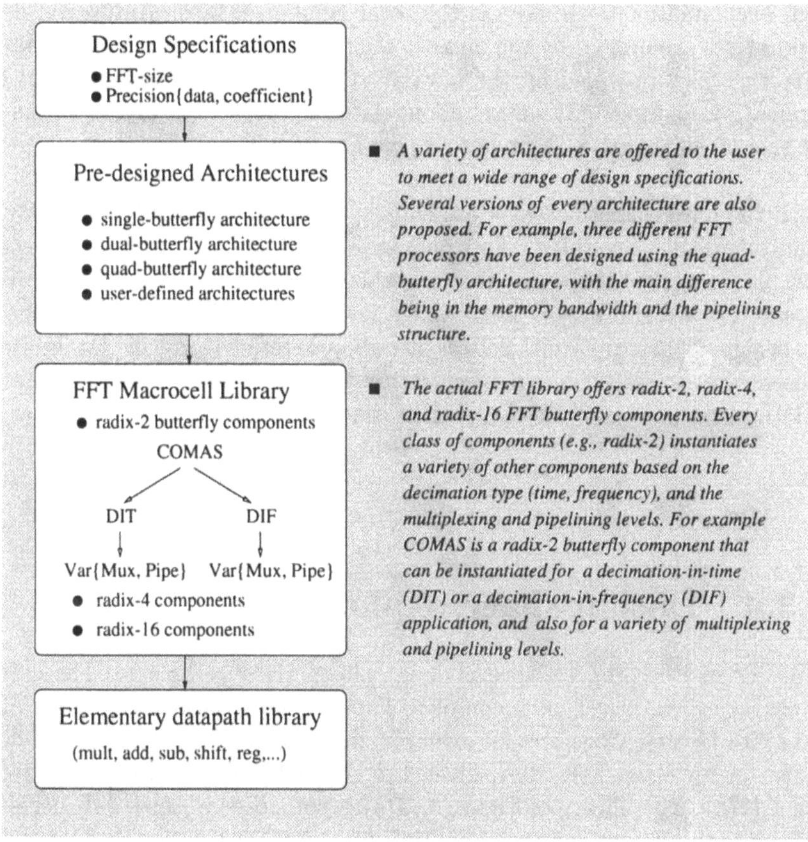


**Figure 4.9** *Design for FFT applications.*

### 4.3.3 Design specification

In order to facilitate automatic IC design for a specific application, the designer must first specify an algorithm that is supported by the automatic design line (ADL). Given a target algorithm, a set of design specifications must be entered into the design tool. For example, the design of programmable FIR filters requires several specifications as generics — data precision, coefficient precision, accumulator precision, and the total number of taps. Automatic design alternatives are limited by the number of application-specific cores and the architecture files provided by the library. The design specification of an FFT application requires information about data and coefficient precision and also the size of the target FFT operation (see Figure 4.9).

Once the design specifications are chosen, a VHDL package "passpara.vhdl" is automatically generated. This package is used to pass the design specification parameters to the committed reusable cores and architecture files. These parameters are also passed to all design levels and to the controller of the target design. "passpara.vhdl" defines global parameters that will instantiate the chosen cores to fit the design specifications. Figure 4.10 illustrates a generated VHDL package from the design specifications illustrated in the Parameter box.

### 4.3.4 Design space exploration

This design step proposes several design alternatives to the user. These design alternatives are based on a complete knowledge of the existing cores of the datapath library. Consider, for example, that the goal is to explore the design space for an 8-tap FIR filter, given the core DfR library illustrated by Figure 4.11(a). The library of Figure 4.11(a) is composed of three FIR cores that define three different degrees of multiplexing. Accurate design decisions can be made at early stages of the design process by means of the precise cost information (area, speed) provided by the library. Design space exploration is also reduced to a limited number of design alternatives illustrated by Figure 4.11(b). If the design alternatives offered are not sufficient to meet the specifications, additional design exploration can be easily obtained by pipelining the datapath. For example, alternatives 4, 5, and 6, illustrated by Figure 4.11(c) are obtained by respectively pipelining alternatives 1, 2, and 3.

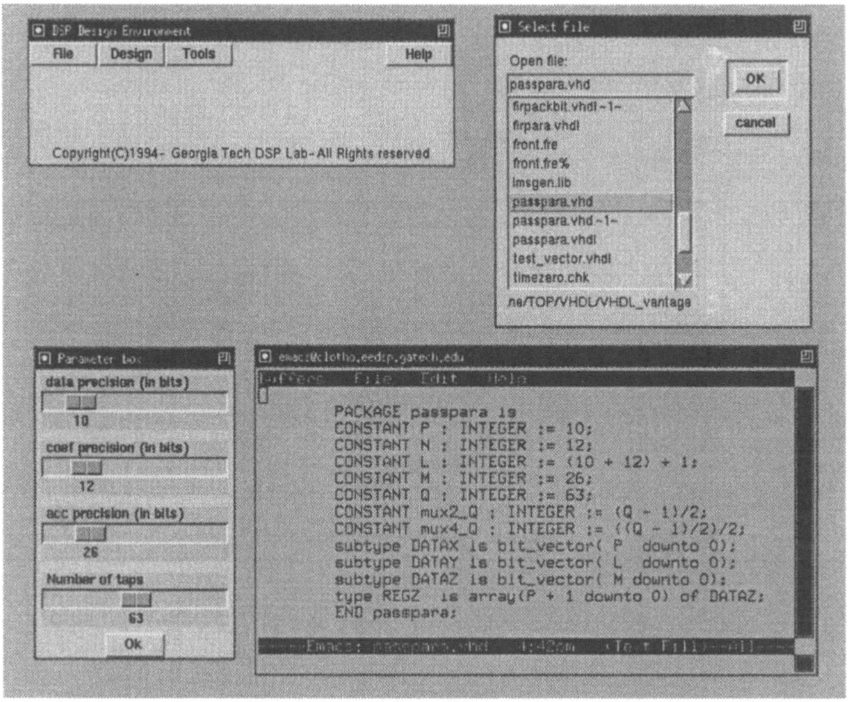


**Figure 4.10** *Invocation of the ADL Design Line.*

Automatic search through the design space can easily be handled by the design tool as a result of the limited but realistic number of alternatives offered. Once, a design alternative is chosen, the appropriate cores are selected and instantiated to fit the design specifications.

For instance, when designing an FIR filter, several degrees of multiplexing and levels of pipelining are supported. The user is invited to commit her design to a unique alternative, given the on-line information about area, timing, and power dissipation involved with every alternative. The design of FFT processors includes design alternatives that specify: the number of butterflies to be used, their radix, multiplexing degree, and pipelining level. Once a design alternative has been specified, the design environment selects from the library the appropriate components that will be used to build-up the datapath core, and tags the design to one of the proposed default architectures. The compilation

| RT-Components | Delay (ns) | Area (mm$^2$) |
|---|---|---|
| FIR-M1 | 30 | 0.88 |
| FIR-M2 | 60 | 1.03 |
| FIR-M4 | 120 | 1.61 |

(a)

| Allocation | FIR-M1 | FIR-M2 | FIR-M3 |
|---|---|---|---|
| Alternative 1 | 8 | 0 | 0 |
| Alternative 2 | 0 | 4 | 0 |
| Alternative 3 | 0 | 0 | 2 |

(b)

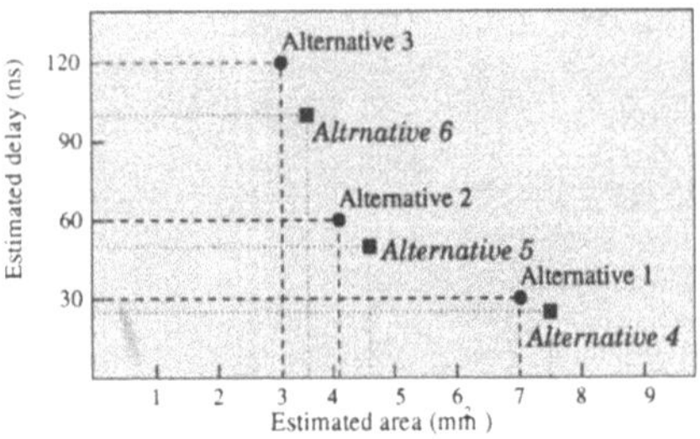


(c)

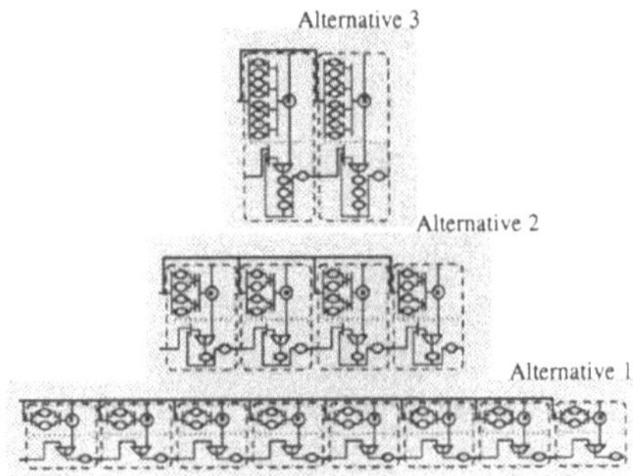


(d)

**Figure 4.11** *(a) An FIR library composed of three datapath cores of different degrees of multiplexing. (b) Resource allocation for an 8-tap FIR filter using the proposed library in (a). (c) Area-delay tradeoffs for different design alternatives. (d) Schematics for alternatives 1, 2, and 3, of the 8-tap FIR filter.*

box shown in Figure 4.12 illustrates the different degrees of multiplexing and levels of pipelining supported for FIR filter design.

### 4.3.5 DfR core generation

Once the appropriate design alternative is chosen and the abstract methods are hooked to the appropriate reusable cores, VHDL code for the entire design is automatically generated. A hierarchical tree-based approach is followed to generate the code. Based on the specified design parameters, a top level VHDL code calls up VHDL programs that themselves make calls to other VHDL cores. These cores are instantiated using the design parameters specified by the user through the previously generated VHDL package "passpara.vhdl". The generated code is composed of three main programs: (1) The datapath, (2) The Controller, and (3) VHDL test vectors for testing the datapath, the control, and the entire design. The user can also input his own VHDL test vectors. Our design environment utilized Viewlogic simulation tools for design verification. However, any VHDL simulation tool supporting the 1076-1987/1993 VHDL standard [?] can be used instead. Figure 4.14 illustrates Vantage simulation of the generated FIR filter with the specifications illustrated in the parameter and compilation boxes.

### 4.3.6 Synthesis

The final step of the automatic design process involves the synthesis of the target design using one of many commercially available design automation synthesis tools. COMPASS synthesis tools were adopted by our design environment. The datapath is synthesized bottom-up using our reusable datapath library. The controller is synthesized top-down in random logic. Test vectors are also generated for logic/gate level simulation of the synthesized design.

*Synthesis of the Datapath* — The datapath is synthesized bottom-up using the DfR library. That is, the datapath of the design is assembled using the pre-designed interoperable cores defined by the datapath library. Resources are allocated from the macro-cell library and tagged to a specific architecture using the provided architectural files. The tradeoffs between cost and performance are easily drawn and the design exploration space is quickly narrowed

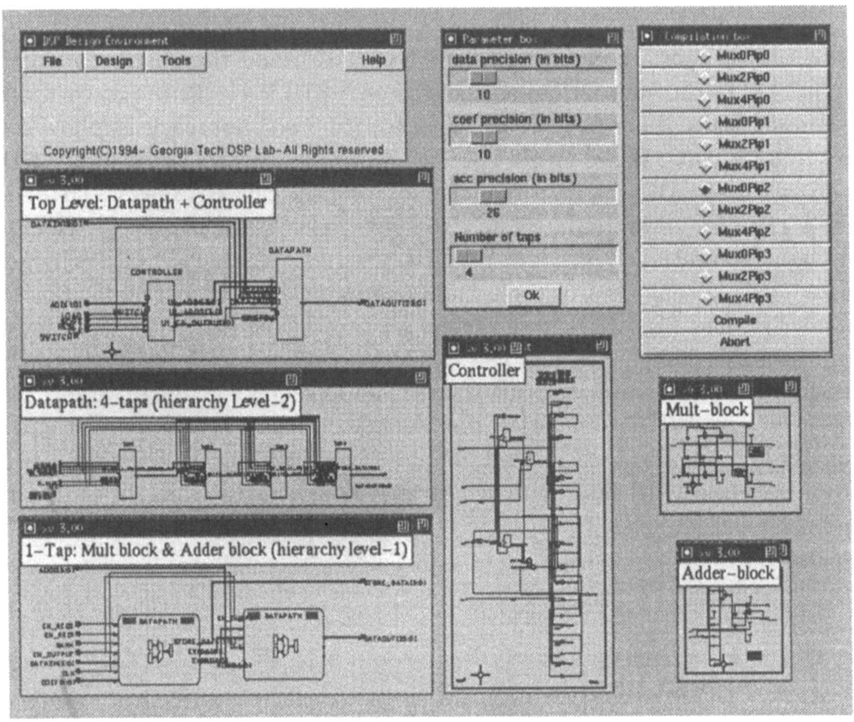


**Figure 4.12** *Synthesis of a 4 Tap FIR filter.*

to a manageable level. Bottom-up design decisions are made with exact cost information provided by the online information about the library cores.

*Synthesis of the Controller* — The controller is synthesized in a top-down manner. A logic synthesizer was used for this purpose. In contrast to datapath structures, control logic is best synthesized using a top-down design methodology. The inherent single-bit structure of control logic makes every logic operation a unique synthesizable operation that maps perfectly to random logic. Additional benefits result from the maturity of automatic optimization of control logic.

Figure 4.12 illustrates the result of the synthesis of an FIR filter with the specifications shown in the parameter box. The hierarchical design approach is illustrated here. At the topmost level we find the complete design (Controller, Datapath). The datapath is a customized architecture file that uses macro-cell components from the datapath library. A single core (an FIR tap) is also illustrated, and is composed of datapath sub-blocks with varying precision. The controller is the result of random logic synthesis.

## 4.4 SIMULATION DESIGN VIEW

The simulation design task is concerned with the verification of the functionality and timing of the target design at all abstraction levels. Within the reuse-based environment proposed, the bulk of the verification process is done in advance (during the DfR cycle). Reusable libraries can slash a substantial amount of time from the verification process. The simulation of the design as a whole is still necessary, but there is a big saving with respect to the individual pre-designed pre-verified cores. The saving is even bigger when reuse is considered at a larger scale (i.e., the functional level). In that case the designer is only required to manipulate a small number of pre-verified components and to verify their interconnectivity to deliver the right function.

The cores provided by the library are simulated at all design levels. A VHDL simulator was used for simulating these library components at high abstraction levels. COMPASS simulation tools (Sim, Qsim, and Qtv) are adopted at the gate and layout levels (synthesized designs can be simulated using any other logic/gate timing simulator).

Designs that are synthesized using the ADL tool are correct by construction. This is because not only are the cores pre-verified but also the generic architectures. Test vectors are also automatically generated to verify the functionality and timing of these designs at different abstraction levels. The outcome of the simulation might require, in the worst case, addition of buffers to some signals with increasing load (i.e., for high order FIR filters. Input signals, such as *data input* and *coefficient input*, drive a large number of cells and require additional buffering).

When using the FDL tool, only the functionality of the library cores is guaranteed and can be verified via provided test vectors. The user is invited to input his own test vectors in order to verify the functionality of the designed ASIC.

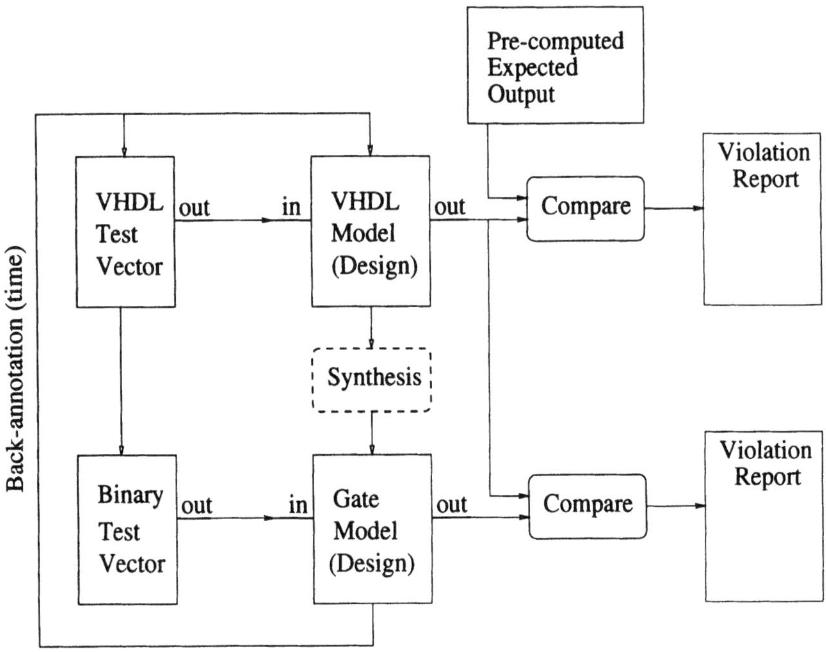


**Figure 4.13** *Simulation procedure for functionality correctness.*

The procedure followed in testing the functionality of designed circuits is illustrated by Figure 4.13. A test vector is hooked to the input pins of the design to be tested. Driven by the test bench, the simulation runs and the output of the VHDL model is compared with a pre-computed vector of expected outputs (given the input test vector). The result of the comparison process is a report that reflects the violations with respect to the expected output. The design is corrected and the simulation is resumed until no violations are recorded. The end of this simulation phase guarantees the correct functionality (to a certain extent defined by the efficiency of the test vectors) of the VHDL model but does not guarantee the correct functionality of the design at the gate level. The bulk of the simulation process is, however, already completed at the functional level.

The simulation procedure at the gate level is then undertaken (see Figure 4.13), but there appears little need to run very long and time consuming gate-level simulations, as has been in the past. Instead, the gate-level simulation will just have to validate (quickly) the higher level simulation. The outcome of the simulation at the annotated behavioral level and and the gate level is compared

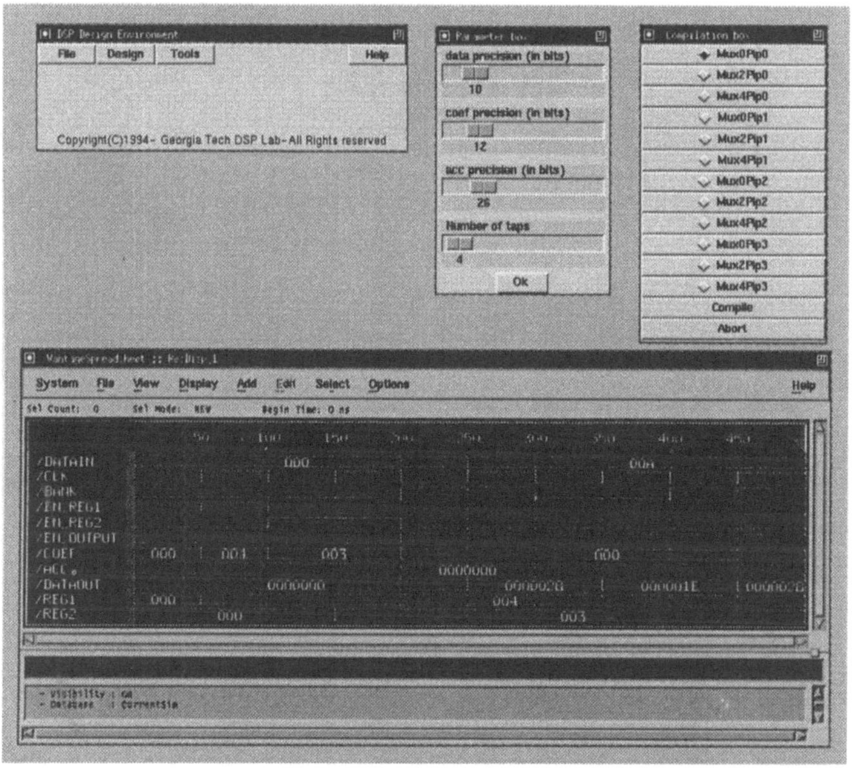


**Figure 4.14** *Design space exploration user interface.*

to detect violations due to improper functionality or a timing mismatch. Back-annotation of essential gate-level timing into the VHDL model is required in order for the simulation at both levels to match. The testing procedure described by Figure 4.13 was handled manually.

The choice of the test vectors is important in increasing the likelihood of full functional correctness. It is recommended that pre-designed and parameterized (reusable) test vectors be supplied with the DfR core library. Such test vectors perform functionality validation that reflects typical application-specific tests. For example, FIR filters are tested for both impulse and step responses. All states of the controller should also be exercised.

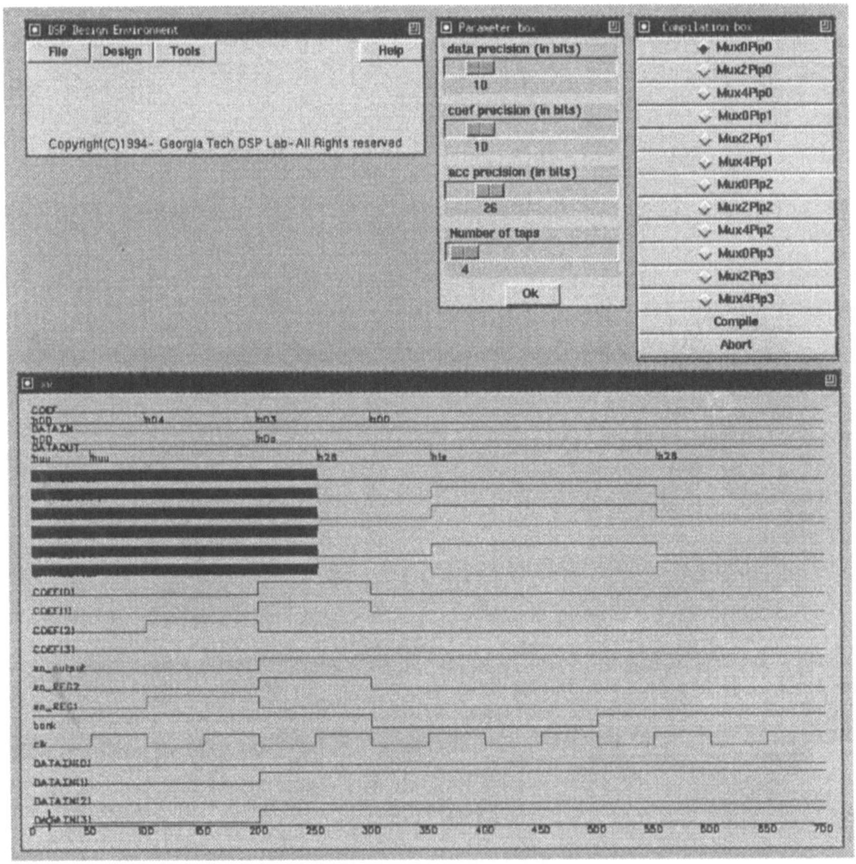


**Figure 4.15** *Design space exploration and logic/gate simulation after synthesis.*

Figure 4.14 illustrates the gate level simulation of the synthesized core-based FIR filter with the specifications in the parameter and compilation boxes. Immediate verification of high level simulations with the corresponding gate/layout level simulation is possible through the application of the generated test vectors.

The logic/gate simulation of synthesized circuits is compared with the high-level simulation of the original VHDL code. Note the conformity of the simulation of the design at both VHDL and gate level simulation (See Figures 4.14 and 4.14,

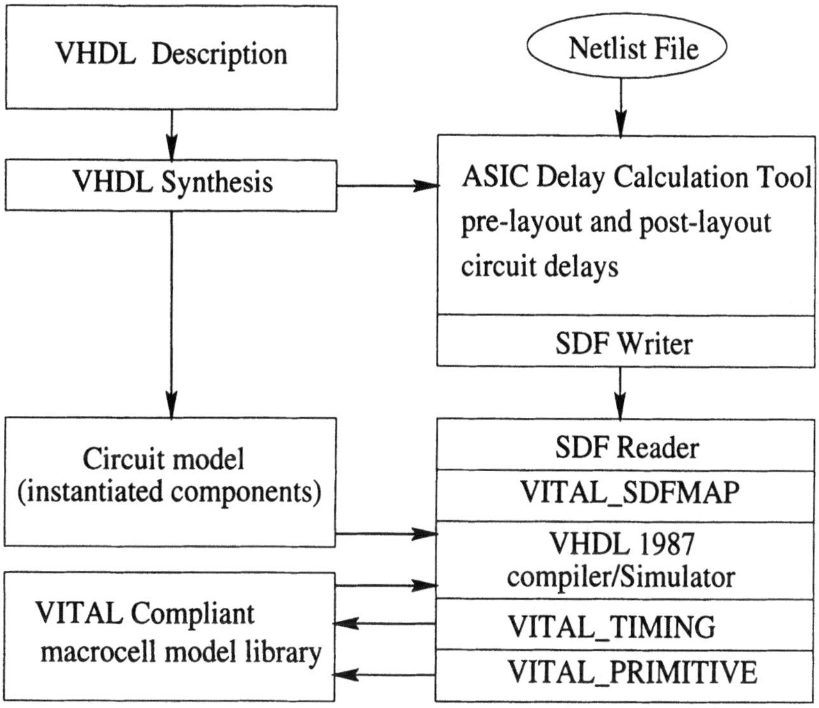


**Figure 4.16** *Automatic back-annotation of VITAL-compliant models*

i.e., at time $t = 250ns$ the output of the filter is the hexadecimal value $h2B$. When switching the working coefficient bank, the FIR filter uses the shadow coefficient bank and the output of the filter changes to $h1B$).

The industry-led VITAL (VHDL Initiative Towards ASIC Libraries) provides accurate timing information for performing sign off simulation of ASICs [74, 75] quickly. VITAL proposes systematic back-annotation of timing characteristics of synthesized layout into high level simulation models [76]. These VHDL models are called VITAL-compliant models. The timing is passed to these cores through a Standard Delay File (SDF) that is either provided by the vendor or generated using the appropriate ASIC delay calculation tool. The back-annotation procedure into the VITAL-compliant models is illustrated by Figure 4.16 [74].

# 4.5 COST MODEL

In this section we model the cost of FASICs (FASICs are the ASICs designed within the functional reuse environment proposed) based on the design attributes and product forecast listed in Table 2.1. The data is typical for a 10,000-gate design, according to research by Dataquest and Integrated Circuit Engineering (ICE) quoted in Liu's work. A cost comparison between FASICs, ASICs, and FPGAs is carried out, based on the derivation of a break-even analysis. The break-even point is quantified in terms of product volume over which one alternative is considered more profitable than the other. The advantages and limitations of the different design approaches considered in this section are listed in Table 4.18.

## 4.5.1 Cost and Time Attributes for FASICs

Tables 4.1 and 4.2 list typical cost and time attributes of a 10,000-gate design, respectively. These attributes are provided for the different design methodologies of interest. The attributes of the ASIC and the FPGA approaches are taken from [46], while those of the FASIC alternative are derived from the several chips built using the design lines proposed in this chapter — ADL and FDL.

### *Time attributes for FASICs*

The time attributes of the FASIC approach are illustrated in Table 4.2. The design time (not including fabrication time) for every design alternative is obtained by adding items #17 through #23 from Table 4.2.

$$Design\_time\_ASIC = \sum_{i=17}^{23} ASIC\_Item(i) = 16\ weeks,$$

$$Design\_time\_FASIC\_FDL = \sum_{i=17}^{23} FASIC\_FDL\_Item(i) = 4\ weeks,$$

The time attributes of the FASIC-ADL (a turn-key approach) are measured by the hour. The sum of items #17 through #23 for the ADL approach is estimated to one day.

$$Design\_time\_FASIC\_ADL = \sum_{i=17}^{23} FASIC\_ADL\_Item(i) = 1\ day.$$

For this specific example we have:

$$\frac{Design\_time\_ASIC}{Design\_time\_FASIC\_FDL} = 4,$$

$$\frac{Design\_time\_ASIC}{Design\_time\_FASIC\_ADL} = 112.$$

Hence the FASIC-FDL approach cuts the design time of ASICs by approximately a factor of 4. The FASIC-ADL approach, on the other hand, cuts the design of ASIC by approximately a factor of 100. These results were also verified during the design of many chips with the proposed approaches.

## *Cost attributes for FASICs*

The cost attributes for FASICs are listed in Table 4.1. The engineering cost, the NRE cost, and the tool cost for the FASIC approach are derived as follows.

1. *Engineering cost* — The engineering cost is proportional to the design time. Since FASIC-FDL cuts the design time of ASICs by approximately a factor of 4, the engineering cost is also cut by approximately the same factor. The engineering cost for the FASIC-ADL approach is cut by approximately a factor of 100.

$$Engineering\_costs\_FASIC\_FDL = \frac{Engineering\_costs\_ASIC}{4}, \quad (4.1)$$

$$Engineering\_costs\_FASIC\_ADL = \frac{Engineering\_costs\_ASIC}{112}. \quad (4.2)$$

   Equations 4.1, and 4.2 are used to derive item #12 in Table 4.1 for FASIC-FDL and FASIC-ADL, respectively.

2. *Non-Recurrent Engineering (NRE)* — The NRE cost incurred by traditional ASICs is related to the fabrication of a hardware prototype for simulation and testing. FASICs rely on VHDL ASIC simulation and VHDL

**Table 4.1** Cost Attributes for FASICs, ASICs, and FPGAs.

| Cost Attributes | FASIC | | ASIC | FPGA |
|---|---|---|---|---|
| | ADL | FDL | | |
| 12. Engineering costs | \$1000 | \$20,000 | \$79,000 | \$25,000 |
| 13. NRE | \$0 | \$0 | \$25,000 | \$0 |
| 14. tools | \$20,000 | \$15,000 | \$10,000 | \$10,000 |
| 15. Average price | \$13 | \$13 | \$13 | \$39 |

system level simulation and integration (see chapter 6). Hence, no hardware prototype is made before validation of the ASIC design. As a result, FASICs yield no NRE cost overhead.

3. *Tools* — FASICs yield an additional tool-related cost overhead over ASICs. FASIC-ADL's additional cost overhead is related to the automation of the design and the use of reusable blocks. This overhead is estimated to \$10,000. FASIC-FDL's (tool-related) cost overhead over ASICs covers the cost of the functional reusable library. It is estimated to \$5,000.

## 4.5.2 Basic break-even analysis

The basic break-even analysis does not take into consideration the time-to-market factor. It only considers the NRE cost, the engineering cost, and the unit price. The basic break-even analysis formula, as defined in [46], is:

$$Basic\_break\_even = NRE + engineering + tools + (units * price). \quad (4.3)$$

The basic break-even equations (4.4, 4.5, 4.6, 4.7) for the different design approaches considered here, are derived by using the data in Table 4.1 and Equation 4.3:

$$FPGA = \$25,000\ eng. + \$0\ NRE + \$10,000\ tools + (X\ units * \$39), \quad (4.4)$$

$$ASIC = \$79,000\,eng. + \$25,000\,NRE + \$10,000\,tools + (X\ units * \$13), \quad (4.5)$$

$$FASIC\_FDL = \$20,000\ eng. + \$0\ NRE + \$15,000\ tools + (X\ units * \$13), \quad (4.6)$$

$$FASIC\_ADL = \$1,000\,eng. + \$0\,NRE + \$20,000\,tools + (X\ units * \$13). \quad (4.7)$$

Solving for the break-even point:

$$X_{ASIC} = 3038\ units,$$

$$X_{FASIC\_FDL} = 0\ units,$$

$$X_{FASIC\_ADL} = -538\ units.$$

These results imply that the ASIC approach is more cost effective than the FPGA approach when the product volume is larger than 3038 *units*. The FDL and ADL approaches, on the other hand, yield better benefits than the FPGA approach for any product volume. These results, however, do not take in consideration the time-to-market constraint. In order to take in consideration this important factor, the analysis should quantify the lost revenue due to a late product delivery.

### 4.5.3 Time-to-market issues

A considerable loss in the revenue might result from late product delivery. This loss is a function of the product lifetime and the delay in the product delivery. Figure 2.10 illustrates the triangular model, developed by Logic Automation, that illustrates the effect of late product delivery on the total product revenue.

The percentage of lost revenue can be derived from this model as illustrated by Equation 2.7. The percentage of lost revenue is relative to the net profit that is formulated by Equation 2.8.

The total development time (design + fabrication) of the different design alternatives consists of adding items #17 through #26 in Table 4.2. The delay $D$ for each alternative is calculated as the difference in turnaround cycles (in

**Table 4.2** Time Attributes for FASICs, ASICs, and FPGAs.

| Time Attributes | FASIC | | ASIC | FPGA |
|---|---|---|---|---|
| | ADL | FDL | | |
| 16. Engineering labor in weeks | - | - | - | 4 |
| 17. Training in weeks | - | 2/7 | 2 | 1 |
| 18. Design capture in weeks | - | 1/7 | 3 | 2 |
| 19. Simulation in weeks | - | 1 | 2 | 2 |
| 20. Test-vector development in weeks | - | 1 | 6 | 0 |
| 21. Place & route in weeks | - | 1/2 | 1 | 1 |
| 22. Back annotation in weeks | - | 4/7 | 1 | 0 |
| 23. Final simulation in weeks | - | 1/2 | 1 | 0 |
| 24. Proto cycle time in weeks | 2 | 2 | 2 | 0 |
| 25. Qualification in weeks | 5 | 5 | 5 | 3 |
| 26. Production leadtime in weeks | 9 | 9 | 9 | 2 |

months) between the FPGA approach and the other design approaches (i.e., ASIC, FASIC).

The lost revenues for the different design alternatives are derived, given the net profit, the percentage of lost revenue, and the delay $D$. The results are illustrated in Table 4.4. The break-even analysis with time-to-market consideration investigates two cases. The first case, only considers the design time (no fabrication time), while the second case considers both the design time and the fabrication time. This separation is necessary because FASICs can be used for system level simulation before fabrication (virtual prototyping, see chapter 6).

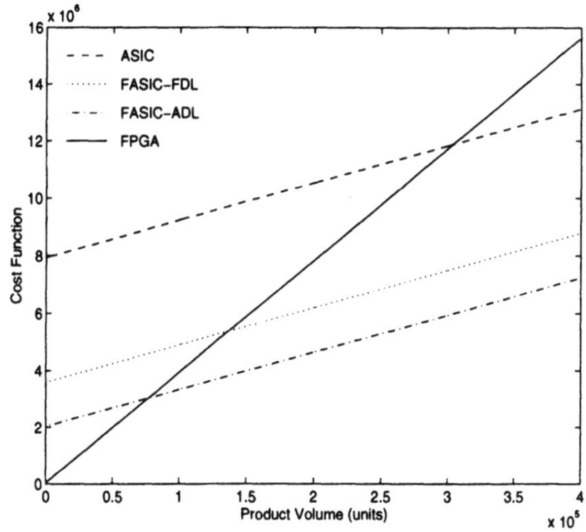


(a)

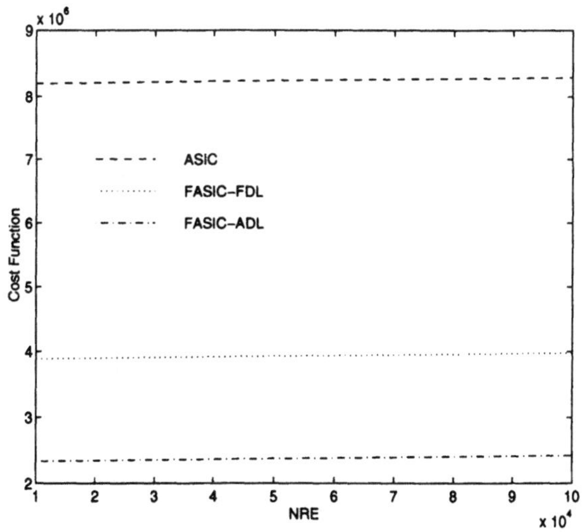


(b)

**Figure 4.17** *(a) Cost functions (b) Effect of NRE.*

The new break-even values are derived by adding the lost revenue (due to delivery delay relative to every approach) to the total cost, and by resolving Equations 4.5, 4.6, and 4.7. The new break-even values are illustrated in Table 4.4. The results show that when the design time (no fabrication time) is included in the cost function, the FASIC approach continues to outperform the FPGA approach, even at very low production volume. When fabrication time is included (here estimated to be 16 weeks [46]), only after the first 76,986 units that the FASIC-ADL approach becomes more profitable than the FPGA approach. The break even for the FASIC-FDL happens at about double that of the FASIC-ADL.

The findings from the previous analysis are:

- The cost of the design is dominated by the time-to-market.
- The fabrication time dominates time-to-market for FASICs.
- FASICs incur no NRE, re-spin, or second source costs.
- The reduction in design turnaround results in the scaling of the engineering costs by the same factor.
- FASIC-FDL and FASIC-ADL cut the design time of ASICs by about a factor of "4X" and "100X", respectively.
- FASICs bridge the gap between ASICs and FPGAs by offering high performances, low cost, and a short time-to-market.

Figure 4.17(a) illustrates the cost function for the different design methodologies when time-to-market is considered. The intersections define the break-even points for the FPGA approach versus the others. Figure 4.17(b) illustrates the effect of adding NRE cost to the total cost function, when time-to-market is considered in the model. In that case, the NRE does not affect the cost function because the model is dominated by the time-to-market cost overhead. Figure 4.18 compares the net revenue corresponding to the different design alternatives considered here.

Table 4.3: Performance comparison between FASICs, ASICs, and FPGAs.

| Issues | FASICs | ASICs | FPGAs |
|---|---|---|---|
| 100k gates plus capability | Yes | Yes | No |
| System integration capability | Yes | limited | limited |
| Special-purpose tool requirement | Yes | No | Yes |
| Sign-off models | Yes | limited | limited |
| Automatic test procedure | Yes | Yes | No |
| Time-to-market | very short-short | long | very short |
| Production volume range | low - very high | high - very high | very low - low |
| Future cost reduction | high | low | high |
| Fabrication re-spin time | No | Yes | No |
| Deep submicron benefits | Yes | Yes | No |
| Speed limit | No | No | Yes |

Table 4.4: Results of the break-even analysis for a 10,000 gate design.

| | Attributes | FASIC | | ASIC | FPGA |
|---|---|---|---|---|---|
| | | ADL | FDL | | |
| no TTM | Break-even | -538 units | 0 units | 3038 units | - |
| with TTM | Development time, weeks | 1/7 (1 day) | 4 | 16 | 11 |
| - | Delay, D, months | - | 1.0 | 4.0 | 2.75 |
| - | Percentage lost revenue | - | 8.18% | 30.86% | 21.75% |
| | Lost revenues due to *D* | - | $1,619,640 | $6,110,280 | $4,306,500 |
| | Break-even | -166,173 units | -103,340 units | 72,415 units | - |
| with TTM | Development time, weeks | 16 | 24 | 32 | 11 |
| - | *D* relative to FPGA, months | 1.25 | 2.25 | 5.25 | - |
| - | Percentage of lost revenue | 10.18% | 17.96% | 39.50% | - |
| - | Lost revenues due to delay *D* | $2,015,640 | $3,556,080 | $7,821,000 | - |
| | Break-even | 76,986 units | 136,770 units | 303,850 units | - |

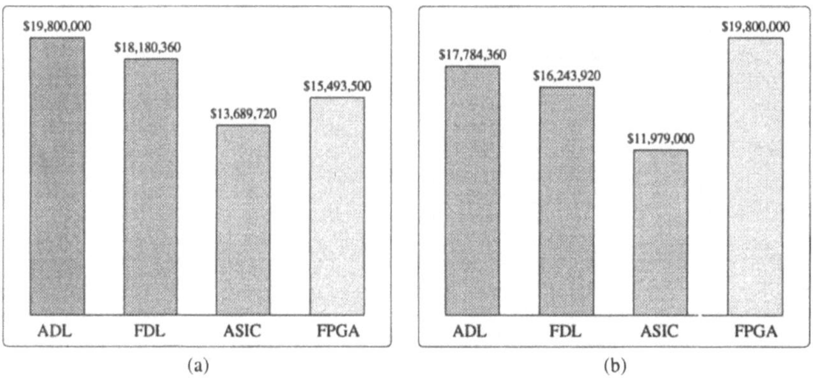


**Figure 4.18** *Relative net revenue for a design with the attributes and product forecast listed in Tables 2.2, 4.1, and 4.2, with time-to-market consideration. (a) only the design time is considered, (b) both design time and fabrication time are considered.*

# 5

# DESIGN WITH REUSE

This chapter describes the core-based design of a few ASICs using the proposed approach of earlier chapters. Applications of interest include a number of signal processors — FIR chips, FFT chips, and related ASICs and ASSPs.

## 5.1 FILTER SYNTHESIS

A synthesizable core DfR library, LMSGEN , has been developed [77] to support the rapid prototyping of digital filters within the design environment described in this monograph. LMSGEN provides generality, flexibility, simplicity, and high performance, at a competitive cost. Generality implies that multiple sample rates, architecture types, and filter orders can be accommodated. Flexibility is offered in terms of a fully-parameterized library in fixed point, and by easy insertion of model-year technology through VHDL. Performance is obtained by providing optimized cores, that implement different degrees of pipelining and several degrees of multiplexing. Power dissipation can be reduced by decreasing the power supply voltage, and taking advantage of the multiple parallel functional units provided. Automatic synthesis of digital filters is supported through the ADL environment.

Prior work that has proposed systematic methods to automated design digital filters is related to FIRGEN [49] that provides ease-of-use at the expense of limited flexibility. Unlike FIRGEN, the systematic design of digital filters through the proposed ADL tool provides for both ease-of-use and flexibility. A wide design exploration capability is assured through a rich and flexible library of cores and architectures.

LMSGEN provides, a programmable FIR filter core library, an LMS (Least-mean squares adaptive filtering) core library, and a Block-LMS core library.

### 5.1.1 The FIR core library

The programmable FIR filter library is characterized by:

- *Programmable FIR-taps and double buffered coefficient bank* — Coefficients can be loaded into one of two reserved banks. One coefficient bank can be written to a "shadow" register, while the other is utilized in the filtering operation. This feature is efficient in adaptive filtering applications. The FIR filter can be combined with an adaptive block, such that the coefficients are updated without interruption in the filtering procedure.
- *Data representation* — Data can be represented internally in both 2's complement and unsigned formats.
- *Parameterized Precision* — Precision of the different datapath cores is parameterized and is passed to the components through a VHDL package (*passpara.vhdl*).
- *Datapath* — Extensive use of optimized datapath cores for precision in excess of 8 bits. Parts of the design that require different precision are implemented in different datapath cores.
- *Pipelining* — Different levels of pipelining are supported to meet the different critical path requirements for a variety of applications. Pipelining can be enforced at the level of inter-primitive cells (i.e., pipelining between the multiplier and the adder), or intra-primitive cells (inside the multiplier).
- *Multiplexing* — Different degrees of time multiplexing are supported to accommodate multiple input rates, while keeping the silicon area low.
- *Power minimization* — Multiple parallel functional units provided allows customizing of the power, supply voltage, and throughput tradeoff.
- *Simple control*— The state machine that controls the FIR basic cells is trivial, and incorporation of higher ordering filtering operations is straightforward.

For example, the subset of the FIR library illustrated by Figure 5.1.1 covers FIR applications with data rates up to $70Mbauds$. Cores defined in Figure 5.1.1 are identified with regard to the representation, *(Domain, Structure, Feature)*, defined in the previous chapter. The domain $D$ is fixed to the FIR algorithm ($D \equiv FIR$). The structure $S$ is tagged to the transpose form. The feature $F$ is varied to encompass different levels of pipelining and degrees of multiplexing. Area, power dissipation, frequency of operation, and maximum data rates for the different cores are also given. Area and power estimation for higher order filter constructs using the cores of the datapath DfR library are also provided.

The basic functional core of a transpose-form FIR filter is illustrated by Figure 5.1(a). The implementation into two datapath sub-blocks is essential to accommodate the difference in precision between the multiplication and the accumulation cores. Figures 5.1(b), and 5.2(a) illustrate the "multiplexed-by-two" and "multiplexed-by-four" transpose-form FIR cores, respectively. The transpose-form FIR structure is well-suited for parallel implementations targeted for high speed applications.

Figure 5.3 illustrates the cascading and the functionality of a multiplexed-by-two 4-tap FIR filter. Here registers $REG3$, and $REG6$ are enabled for a write operation once every two clock cycles. The functionality of a multiplexed by four 4-tap FIR filter is also illustrated by Figure 5.4, where register $REG4$ is enabled for a write operation once every 4 clock cycles. In the general case of a multiplexed by $M$ structure, the register that concatenates the elementary cells together is enabled for a write operation once every $M$ clock cycles.

The control for these filters is very simple because of the dataflow nature of their structure. The filter is primarily in one of 3 states — loading the coefficients, filtering, or both. A switching-set key allows the filter to switch from the working register bank to the "shadow" register bank without interruption in its operation. This feature is important for filtering with time varying coefficients. The structure of the controller of Figure 5.2(b) does not change with a variation of the filter's order. The only additional requirement is that the circuit will be driving more cores. Hence appropriate buffering is needed to compensate for the increased load capacitance. The order of the filter is introduced as design parameter, and the ADL tool automatically generates the appropriate control for the filter.

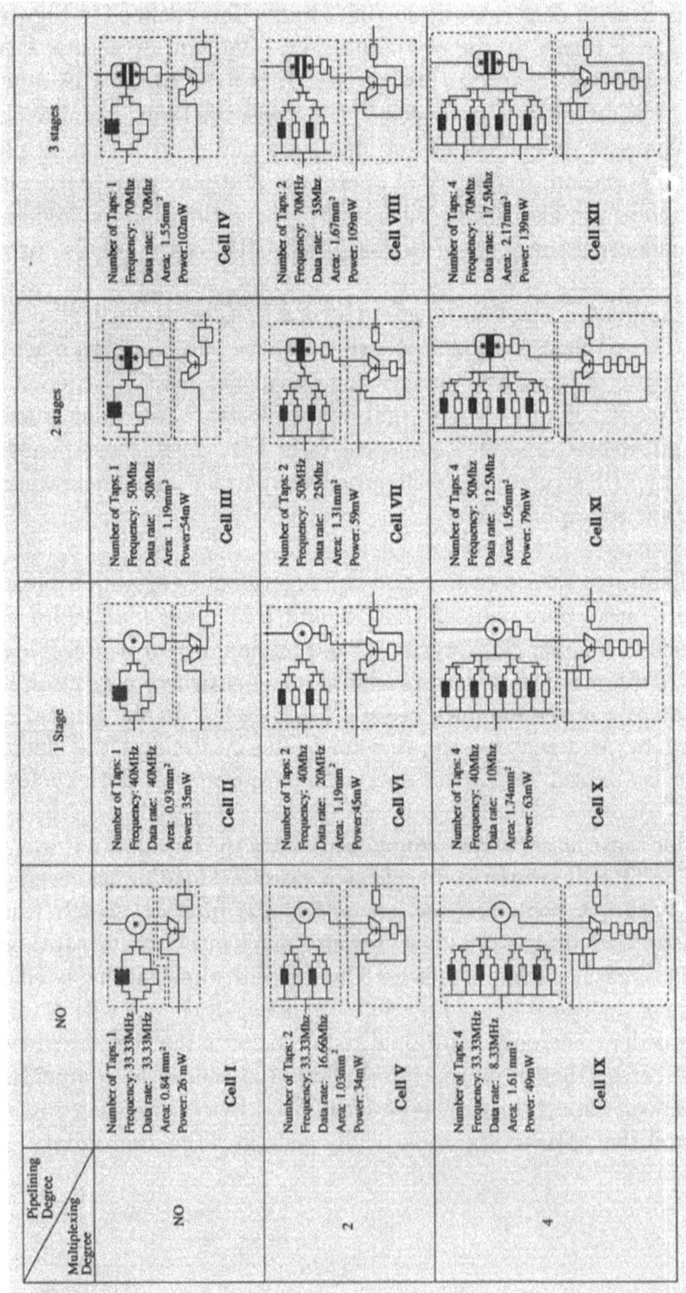


Figure 5.1: *A subset of the FIR library*

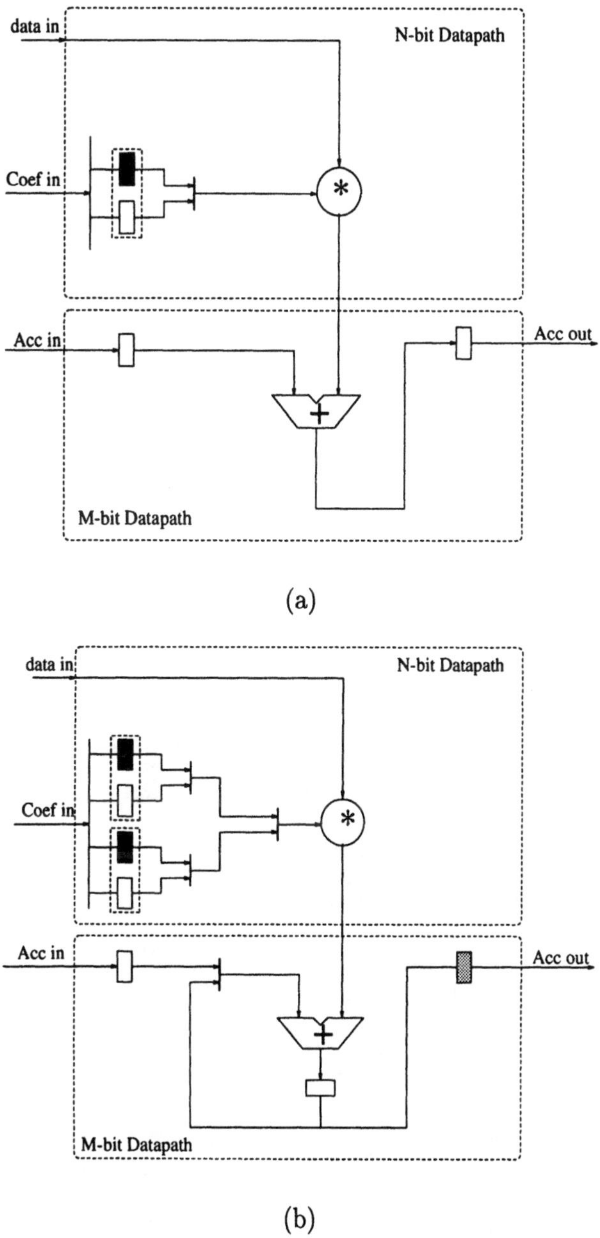


**Figure 5.1** *(a) Datapath of a transpose form FIR tap. (b) Datapath of a multiplexed by 2 FIR core.*

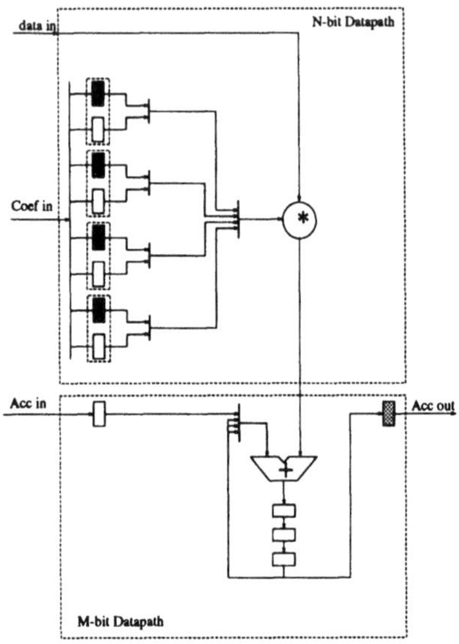


(a)

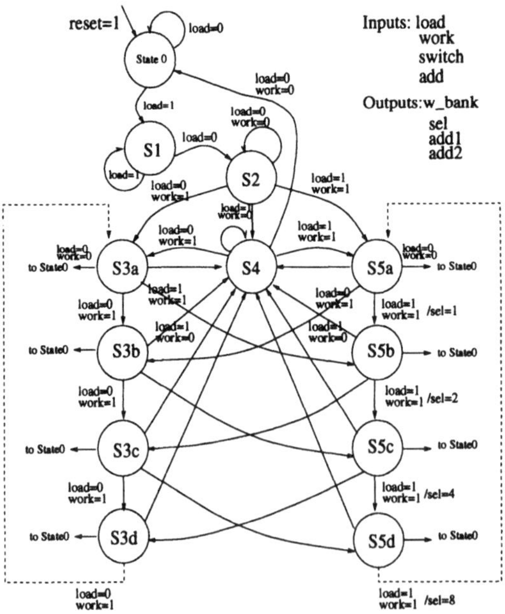


(b)

**Figure 5.2** *(a) Datapath FIR core. (b) Controller.*

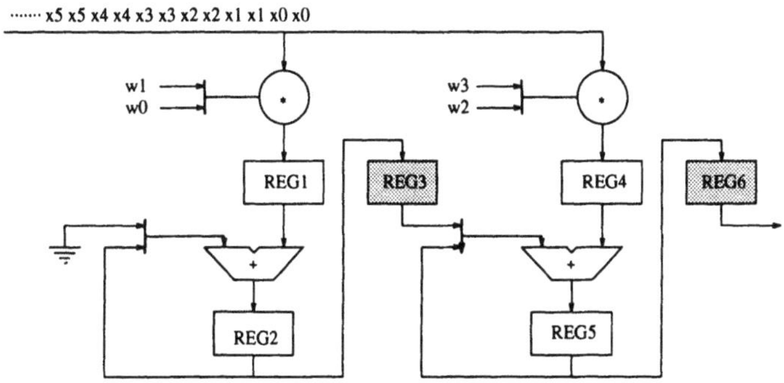


| CLK | REG1 | REG2 | REG3 | REG4 | REG5 | REG6 |
|---|---|---|---|---|---|---|
| 1 | w1*x0 | — | — | w3*x0 | — | — |
| 2 | w0*x0 | w1*x0 | — | w2*x0 | w3*x0 | — |
| 3 | w1*x1 | w0*x0 | w1*x0 | w3*x1 | w2*x0 | w3*x0 |
| 4 | w0*x1 | w0*x0+w1*x1 | w0*x0 | w2*x1 | w2*x0+w3*x1 | |
| 5 | w1*x2 | w0*x1 | w0*x0+w1*x1 | w3*x2 | w1*x0+w2*x1 | w2*x0+w3*x1 |
| 6 | w0*x2 | w0*x1+w1*x2 | w0*x0+w1*x1 | w2*x2 | w1*x0+w2*x1<br>+w3*x2 | |
| 7 | w1*x3 | w0*x2 | w0*x1+w1*x2 | w3*x3 | w0*x0+w1*x1<br>w2*x2 | w1*x0+w2*x1<br>+w3*x2 |
| 8 | w0*x3 | w0*x2+w1*x3 | w0*x1+w1*x2 | w2*x3 | w0*x0+w1*x1<br>+<br>w2*x2+w3*x3 | |
| 9 | w1*x4 | w0*x3 | w0*x2+w1*x3 | w3*x4 | w0*x1+w1*x2<br>+ w2*x3 | w0*x0+w1*x1<br>+<br>w2*x2+w3*x3 |
| 10 | w0*x4 | w0*x3+w1*x4 | w0*x2+w1*x3 | w2*x4 | w0*x1+w1*x2<br>+<br>w2*x3+w3*x4 | |
| | | | | | | w0*x1+w1*x2<br>+<br>w2*x3+w3*x4 |

**Figure 5.3** *Multiplexed by two 4-tap FIR architecture and functionality. Registers REG3, and REG6 are enabled for a write operation once every two clock cycles.*

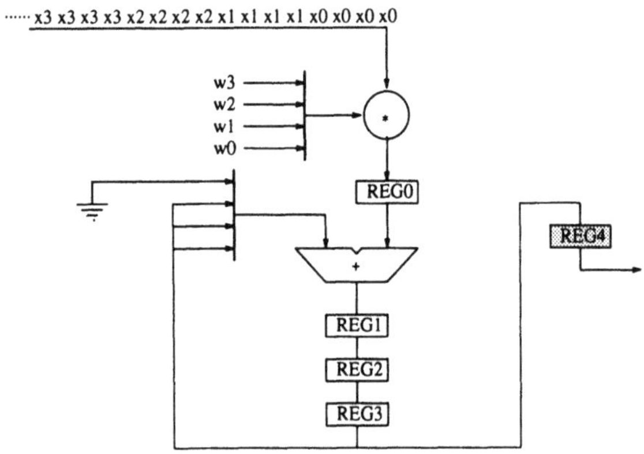


| CLK | REG0 | REG1 | REG2 | REG3 | REG4 |
|---|---|---|---|---|---|
| 1 | w3*x0 | | | | |
| 2 | w2*x0 | | | | |
| 3 | w1*x0 | | | | |
| 4 | w0*x0 | | | | |
| 5 | w3*x1 | w0*x0 | | | |
| 6 | w2*x1 | | w0*x0 | | |
| 7 | w1*x1 | | | w0*x0 | |
| 8 | w0*x1 | w0*x0+w1*x1 | | | w0*x0 |
| 9 | w3*x2 | | w0*x0+w1*x1 | | |
| 10 | w2*x2 | | | w0*x0+w1*x1 | |
| 11 | w1*x2 | w0*x0+w1*x1 +w2*x2 | | | w0*x0+w1*x1 |
| 12 | w0*x2 | | w0*x0+w1*x1 +w2*x2 | | |
| 13 | w3*x3 | | | w0*x0+w1*x1 +w2*x2 | |
| 14 | w2*x3 | w0*x0+w1*x1 +w2*x2+w3*x3 | | | w0*x0+w1*x1 +w2*x2 |
| 15 | w1*x3 | | w0*x0+w1*x1 +w2*x2+w3*x3 | | |
| 16 | w0*x3 | | | w0*x0+w1*x1 +w2*x2+w3*x3 | |
| 17 | w3*x4 | | | | w0*x0+w1*x1 +w2*x2+w3*x3 |

**Figure 5.4** *Multiplexed-by-four 4-tap FIR architecture and functionality. Register REG4 is enabled for a write operation once every 4 clock cycles.*

### 5.1.2 The LMS core library

The adaptive LMS library contains cores for parallel and sequential implementations of the LMS algorithm that can be combined with the programmable FIR filters described earlier.

The LMS adaptive algorithm consists of minimizing the mean square error function, $J_{ms}$, defined by:

$$J_{ms} = E\{e^2(k)\}, \tag{5.1}$$

where $E\{\cdot\}$ is the expectation function, and $e(k)$ is the error function between some desired function, $d(k)$, and the result of some filtering process, $y(k)$. The error function, $e(k)$, is defined by:

$$e(k) = d(k) - y(k). \tag{5.2}$$

For LMS adaptive FIR filters, the filter coefficients $h_j$, for $j = 0, ..., M-1$, are updated with respect to the following equation:

$$h_j(n+1) = h_j(n) + \mu e(n)x(n-1), \tag{5.3}$$

where $x(\cdot)$ is the input to the FIR filter, and $\mu$ is the adaptation constant that determines both the stability and the speed of convergence of the algorithm.

The hardware implementation of high speed LMS adaptive FIR filters is a difficult task since all the coefficients have to be updated during every sample period. According to Equation 5.3, the coefficient update for a $M-$order FIR filter requires the calculation of $2M$ multiplications and $M-1$ additions during every sample period. Practically, the adaptation constant $\mu$ is usually chosen to be a power of 2 constant, such that the multiplication with $\mu$ is implemented with a barrel shifter.

Figure 5.5 illustrates a sequential architecture of a proposed LMS core. The multiplication with the stepsize, $\mu$, of the LMS algorithm is implemented via a barrel shifter ($\mu$ is programmable, and can be set as small as $2^{-15}$). The datapath implements two different sub-blocks of parameterized precision to account for the difference in word-length between the front-end precision and the internal precision. A parameterized (precision and depth) register file is used to store the coefficients in high precision, where one write operation and 2 read operations can be performed during every clock cycle. The number of the coefficients is also set as a parameter.

Aimed at efficient hardware implementation, different modifications to the LMS algorithm have been proposed in the literature[78, 79, 80]. Such algorithms do

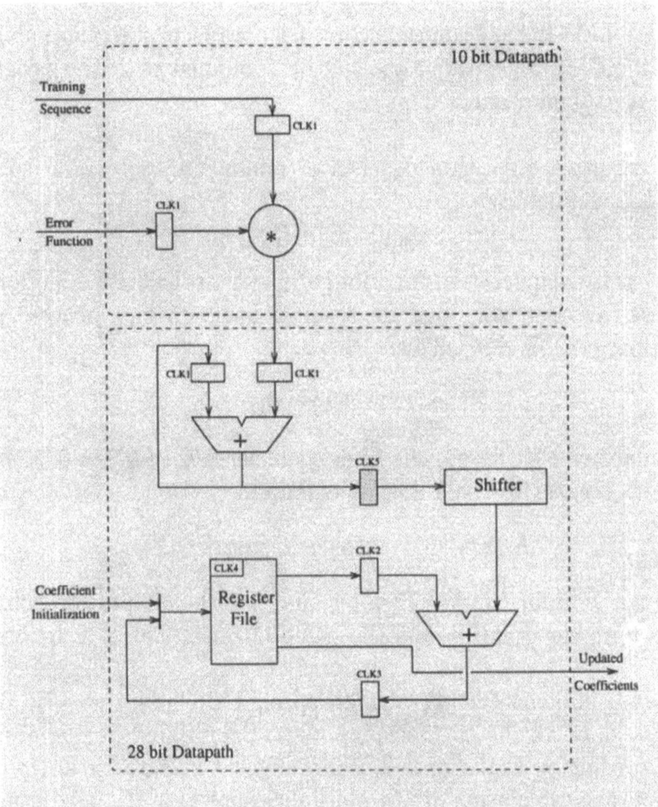


**Figure 5.5** *LMS core (library hierarchy LEVEL-1).*

not perform as well as the original LMS but ease some of the restrictions. The Block-LMS algorithm [81, 82], for instance, was proposed to relax the extensive computational requirements of LMS-adaptive filters. The Block-LMS algorithm updates the coefficients of the filter once every data block input, freeing the hardware from updating the coefficients for each new data input. This approach provides more time for the adaptive system to calculate the new coefficients, and to enable very high speed application at the cost of slower convergence. Slow convergence can be compensated for by the use of training sequences. A training sequence is a well chosen block of data known to both the transmitter and the receiver. It enables a faster convergence of the filter coefficients. Consider the

LMS coefficient update algorithm described by Equation 5.3, it follows that after $N$ coefficient updates we have:

$$h_j(n+N) = h_j(n) + \mu \sum_{i=0}^{N-1} e(n+i)x(n-j+i) \tag{5.4}$$

Which also can be re-written in matrix form as follows:

$$\underline{H}(n+N) = \underline{H}(n) + \mu \underline{\underline{X}}(n)\underline{E}(n) \tag{5.5}$$

Where $\underline{H}(n)$ is the coefficient vector defined by:

$$\underline{H}(n) = [h_0(n)\ h_1(n)\ h_2(n)\ ...\ h_{M-1}(n)]^T, \tag{5.6}$$

$\underline{E}(n)$ is the error vector defined by:

$$\underline{E}(n) = [e(n)\ e(n+1)\ e(n+2)\ ...\ e(n+N-1)]^T, \tag{5.7}$$

and $\underline{\underline{X}}(n)$ is the matrix defined by:

$$\underline{\underline{X}}(n) = \begin{bmatrix} x(n) & x(n+1) & ... & x(n+N-1) \\ x(n-1) & x(n) & ... & x(n+N-2) \\ ... & ... & ... & ... \\ x(n-M+1) & x(n-M+2) & ... & x(n-M+N) \end{bmatrix}$$

Which also can be re-written as:

$$\underline{\underline{X}}(n) = \begin{bmatrix} x(n+M-1) & x(n+M) & ... & x(n+M+N-2) \\ x(n+M-2) & x(n+M-1) & ... & x(n+M+N-3) \\ ... & ... & ... & ... \\ x(n) & x(n+1) & ... & x(n+N-1) \end{bmatrix}$$

The Matrix $\underline{\underline{X}}(n)$ is composed of the pre-chosen data sequence :

$$x(n),\ x(n+1),\ ...,\ x(n+M+N-3),\ x(n+M+N-2), \tag{5.8}$$

which is also called training sequence.

A default architecture for the Block-LMS algorithm [81, 82] is illustrated by Figure 5.6. The datapath core in this architecture is the same as the one illustrated by Figure 5.5. The coefficients, however, are updated only once every $N$ symbol periods ($N$ being the block size). The slower convergence of the Block-LMS algorithm is compensated for by the use of training sequences in video applications. The Block-LMS architecture of Figure 5.6 is composed of two different state machines operating at two different clock frequencies—$f1$ and $f1/N$. The control circuitry is responsible for the following tasks:

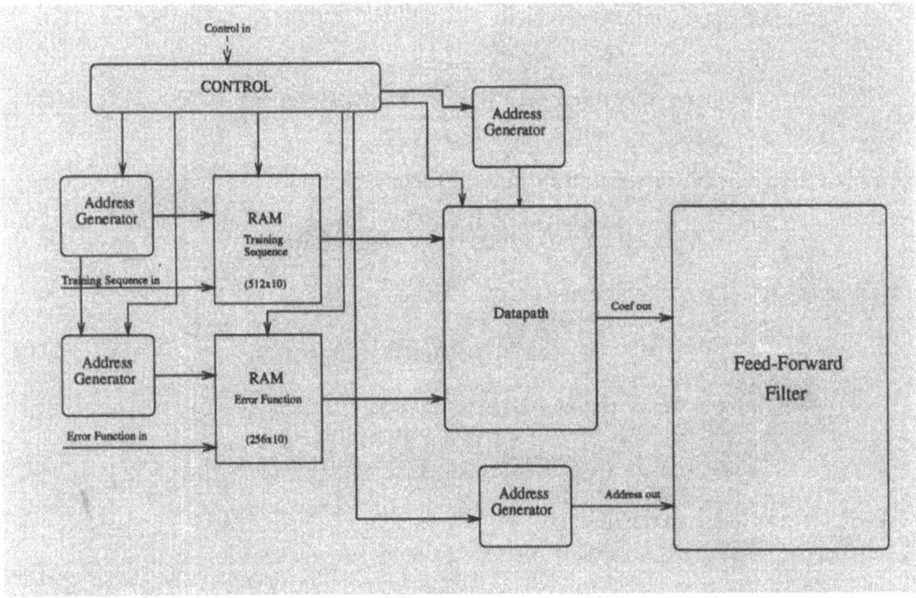


**Figure 5.6** *A Block-LMS core architecture (library hierarchy LEVEL-2).*

- loading the step size,
- loading startup coefficients,
- loading the training sequence into the training sequence RAM,
- loading the error function into the error function RAM,
- updating the coefficients with regard to Equation 5.5,
- loading the coefficients back into the feed-forward filter.

Each step involves the generation of the right addresses for the training sequence RAM, the error function RAM, the register file, and for the external coefficient registers.

### 5.1.3 DwR results

Several high performance digital filters have been designed and tested in design times varying from one day to a few weeks at Georgia Tech DSP Laboratory using the proposed approach. Some of these are described next.

#### *Chip 1 — Block-LMS adaptive Filter*

A 60−tap-Block LMS FIR filter has been designed in $0.8\mu$ CMOS technology and tested for a maximum operation frequency of $50MHz$ and a maximum data rate of $12.5Mbaud$. The core area of this chip is $81.95mm^2$, and its power dissipation is $1.387W$. The floorplan of this implementation is illustrated by Figure 5.8(a). The chip is composed of a feed-forward core, (60-tap programmable FIR filter), and a feedback core that implements the adaptive Block-LMS algorithm.

The feed-forward block of the filter is a 60-tap programmable FIR filter. The design is generated using the ADL tool with the following design specification; 10-bit data and coefficient precision, 26-bit internal precision, and a number of taps $Q = 60$. A VHDL parameter passing package is then generated, and several design alternatives are offered (sequential structure, parallel direct-form structure, and parallel transpose-form structure). The transpose-form structure, with two pipeline stages and a multiplexing degree of 4, is chosen to accommodate a maximum data rate specification of $12.5Mbauds$. The library core **Cell XI** (see Figure 5.1.1) is used by the ADL tool as a building block for the datapath of the filter.

The feedback block adopts the default architecture (library hierarchy LEVEL-2) illustrated by Figure 5.6 where the datapath core is the Block-LMS component as shown in Figure 5.5 (library hierarchy LEVEL-1). The coefficients are updated only once every $N$ symbol periods. With the block size, $N$, fixed to 255, the update of the coefficients is based on the following equation:

$$\begin{bmatrix} h_0(n+255) \\ h_1(n+255) \\ \cdots \\ h_{59}(n+255) \end{bmatrix} =$$

$$\begin{bmatrix} h_0(n) \\ h_1(n) \\ \cdots \\ h_{59}(n) \end{bmatrix} + \mu \begin{bmatrix} x(n+59) & x(n+60) & \ldots & x(n+313) \\ x(n+58) & x(n+59) & \ldots & x(n+312) \\ \cdots & \cdots & \cdots & \cdots \\ x(n) & x(n+1) & \ldots & x(n+59) \end{bmatrix} \begin{bmatrix} e(n) \\ e(n+1) \\ \cdots \\ e(n+59) \end{bmatrix}$$

The block diagram of the address generators and the timing diagram for the Block-LMS architecture are illustrated by Figures 5.7(a), and 5.7(b), respectively.

## *Chip 2 — Programmable FIR Filter*

A 64–tap programmable FIR filter has been designed in $0.8\mu$ CMOS technology and tested for a maximum operation frequency of $50MHz$ and a maximum data rate of $12.5Mbaud$. The core area of this chip is $61.35mm^2$ (total chip area = $90.67mm^2$), with a power dissipation of $1.331W$. The floorplan is illustrated by Figure 5.8(b).

## *Chip 3 — Lattice Filter*

An IIR lattice filter (time scaled by two) has been designed in two-man days, within the FDL environment. The RTL structure of the lattice IIR filter and the library cores (*Cell1_1b* and *Cell1_1b_bis*) used to build the filter are illustrated by Figure 5.9.

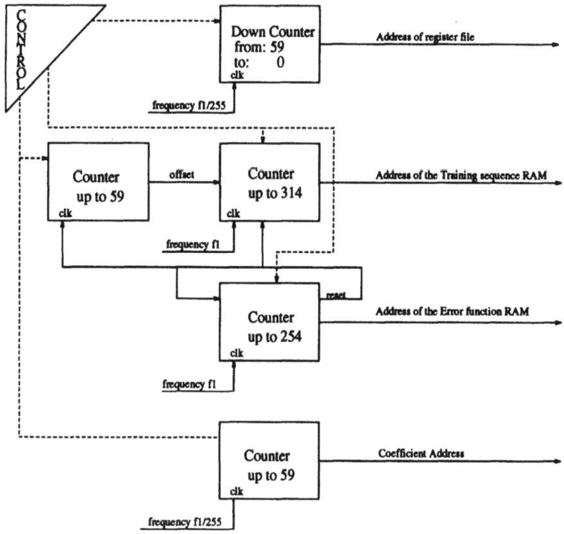


(a)

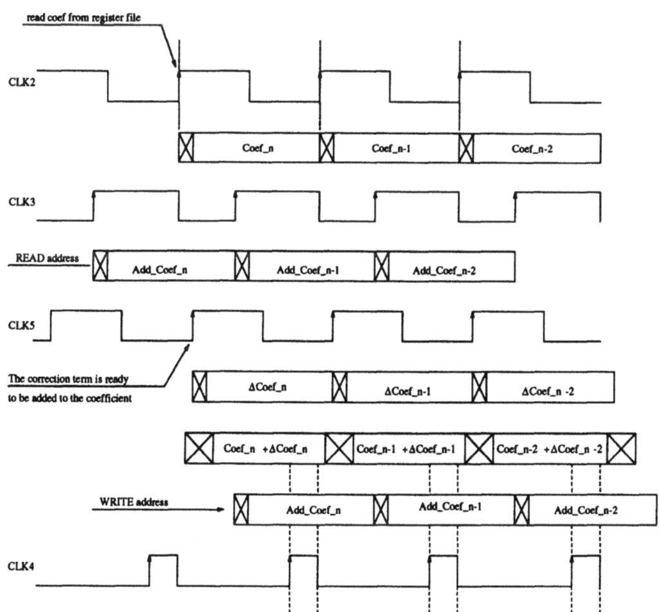


(b)

**Figure 5.7** *(a) Address Generator Core, (b) Timing Block-LMS.*

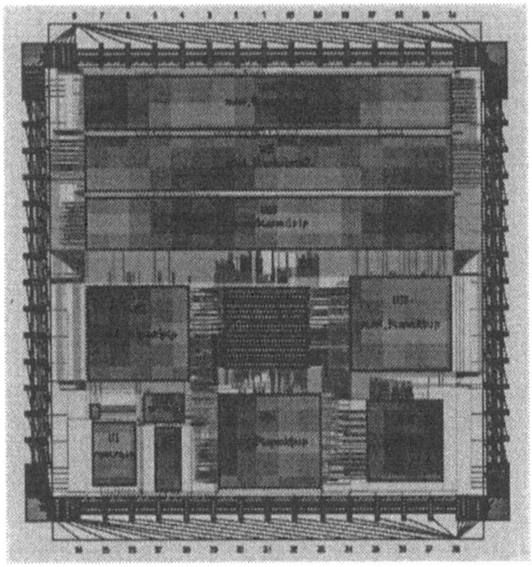

(a)

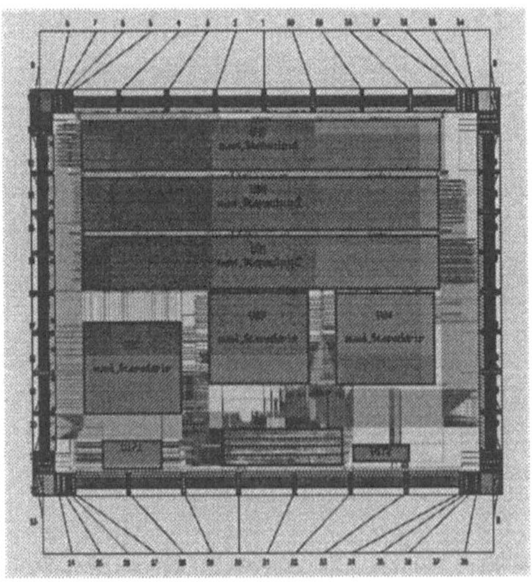

(b)

**Figure 5.8** *(a) 50 Mhz, 1.387 W, 60-tap LMS chip, (b) 50 Mhz, 1.33W, 64-tap FIR chip.*

These cores were selected from the FIR library. An RTL VHDL file using the selected components, *Cell1_1b* and *Cell1_1b_bis*, was written to describe the datapath structure of the lattice filter. Figure 5.10 illustrates the RTL description of the lattice filter using the *generate* statement. The controller for the lattice filter was written at the dataflow level in VHDL and synthesized in a top-down manner. A 5-stage $0.8\mu$ CMOS VLSI implementation of the lattice IIR filter is illustrated by Figure 5.9(b). The filter occupies $15.20mm^2$ of silicon area, and dissipates $0.25W$ of power at a maximum frequency of $25MHz$.

## *Chip 4 — All-Digital Oversampled Sensor (ADS)*

The All-Digital oversampled Sensor (ADS) chip illustrated by Figure 5.13(b) is an implementation of the ADS algorithm proposed in [83]. The same parameterized VHDL model of the ADS can be configured to implement the SCDFE algorithm described in [84]. The ADS algorithm proposes the use of oversampled digital front-end filters instead of the commonly used analog anti-aliasing filters, resulting in a performance/cost-effective DSP application. Analog anti-aliasing filters have a number of disadvantages with respect to cost, power, stability, and ease of integration in VLSI [85, 86, 87].

The structure of the ADS chip is shown in Figure 5.11(a). The sensor oversamples the input signal plus noise, $s(t)$, $(= x(t) + \eta(t))$, at the Nyquist rate of the wide-band white additive noise signal $\eta(t)$, which is many times higher than the Nyquist frequency of the narrowband information signal, $x(t)$. The sampled information and noise sequences are denoted $x[n]$ and $\eta[n]$, respectively. Due to considerable oversampling of $x(t)$, the signal samples $x[n]$ are highly correlated, while the noise samples $\eta[n]$ are not.

The narrowband characteristics of the desired signal allow to estimate $x[n]$ from past samples of the sequence $s[n]$. This is realized by exploiting the high correlation between these two signals. The input to the LMS-FIR block of the sensor is a delayed version of the received signal $s[n]$. A decorrelating delay $D$ is used for this purpose in order to eliminate any correlation between $\eta[n]$ and $\eta[n-D]$. $D$ should not be very large such that the correlation between $x[n]$ and $x[n-D]$ is lost.

Modem simulation results for a 2B1Q baseband signal (with a baud rate of $2KHz$, and a sampling rate of $f_s = 50KHz$) are illustrated by Figures 5.11(b), 5.12(a), (b), and 5.13(a). Figure 5.11(b) illustrates the performance of the

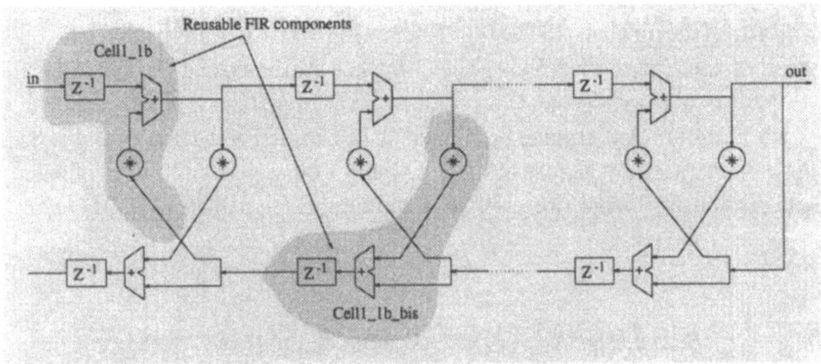


(a) RTL core of IIR lattice.

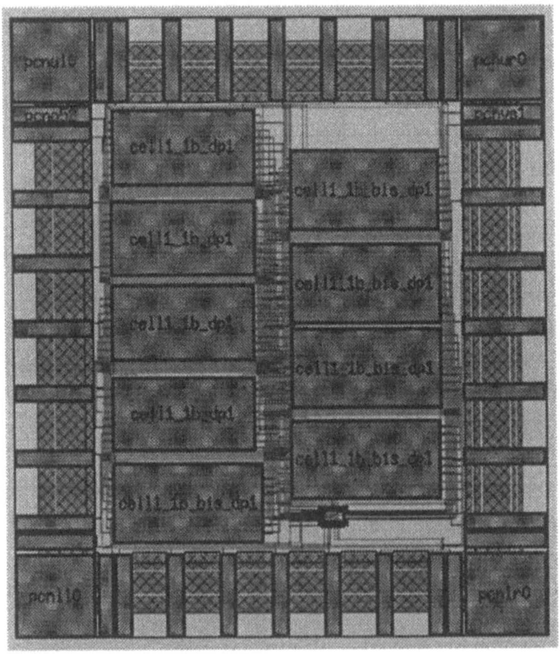


(b) VLSI implementation

**Figure 5.9** *RTL design of an IIR lattice filter structure.*

```
use work.passpara.all;
library ieee;
use ieee.std_logic_1164.all;

entity lattice_iir is
port ( datain : bit_vector(P downto 0);
        clk: in bit;
        add1 : in bit_vector(Q downto 0);
        en_output : in bit;
        coef : in bit_vector(L downto 0);
        dataout : out bit_vector(M down to 0));
end lattice_iir;

architecture gen_lattice_iir of Lattice_iir is
component cell1_1b
        port (datain: in bit_vector (P downto 0);
              clk, en_REG1, en_output : in bit;
              coef : in bot_vector(L downto 0);
              acc : in bit_vector(M downto 0);
              dataout : out bit_vector(M downto 0));
              end component;

component cell1_1b_bis
        port (datain: in bit_vector (P downto 0);
              clk, en_REG1, en_output : in bit;
              coef : in bot_vector(L downto 0);
              acc : in bit_vector(M downto 0);
              dataout : out bit_vector(M downto 0));
end component;

signal accz : REGZ;
signal s: REGZ;
begin
accz(0) <= datain;
G: for i in 0 to Q generate
    cell1_1b_bisi : cell1_1b_bis port map (s(i+1), clk,
            add1(i), en_output, coef, accz(i), accz(i+1));

    cell1_1bi : call1_1b port map (accz(i+1), clk, add1(i),
            en_output, coef, s(i+1), s(i));
end generate;
s(Q+1) <= accz(Q+1);
dataout <= accz(Q+1);
end gen_lattice_iir;
```

**Figure 5.10** *VHDL description of an RTL IIR lattice filter.*

ADS for different number of taps, $L$. Increasing $L$ results in: a steeper rolloff, smaller sidelobes, and better noise cancellation. Figure 5.12(a) compares the performance of the ADS and a digital lowpass filter (LPF) for the same number of taps ($L = 8$). Results show that the ADS has a steeper rolloff, while the LPF results in a much slower rolloff (very wide mainlobe) but a smaller peak sidelobe amplitude. Combining the advantages of these two filters is realized by cascading the ADS and the LPF. This solution results in a front-end filter with narrow transition band (steep rolloff), and a small peak sidelobe amplitude as illustrated by Figure 5.12(a).

Figure 5.12(b) illustrates the results of cascading the ADS and a LPF or a moving-average filter (MAF). The order of the MAF can be as high as the over-sampling factor. Results show that combining the ADS with a MAF provides steeper rolloff and smaller peak sidelobe amplitude. Cascading all three filters (ADS, MAF, and LPF) results in the best performance in terms of rolloff factor and sidelobe amplitude [83]. The best cost/performance compromise, however, is realized by a low order ADS and a MAF. Moving average filters are simple to implement (e.g., multiplier-less), adapt easily to high clock rates, and can be used as decimation filters. A snapshot of the time-domain results for the 2B1Q signal described earlier is illustrated by Figure 5.13(a).

## 5.2 FFT SYNTHESIS

Most general purpose Digital Signal Processors (DSPs) with enhanced FFT performance deliver 5-50 MOPS[1]. These general purpose DSPs are only applicable to a limited range of real-time DSP applications. Several dedicated FFT architectures have been proposed to assist a wider range of real-time DSP applications, with VLSI implementations involving very complex designs and extremely long design turnaround cycles [88, 89, 90].

The SHARP FFT solution [88] proposes a multi-chip architecture that provides real-time performances for a 1024-point FFT under $100\mu s$. Single-chip dedicated FFT processors that deliver real-time performances have also been proposed. O'Brien et al. [89] proposed a 200 MIPS single-chip FFT processor that uses a radix-4 processing scheme. Miyanaga and Yamauchi from NTT LSI laboratories, have proposed a 400 MFLOPS FFT processor [90].

[1] million of operations per second

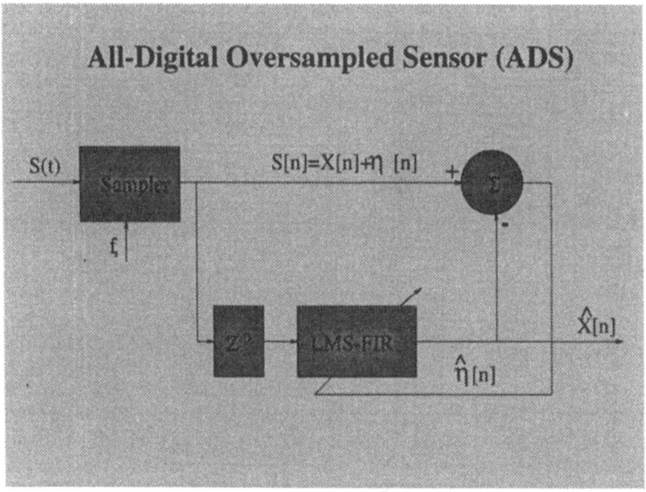


(a)

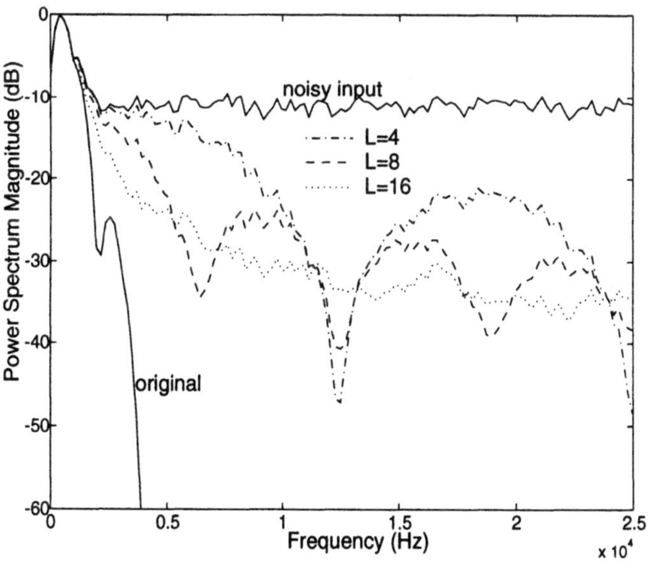


(b)

**Figure 5.11** *(a) Structure of the all-digital sensor (ADS). (b) Performance of the ADS with number of taps, L.*

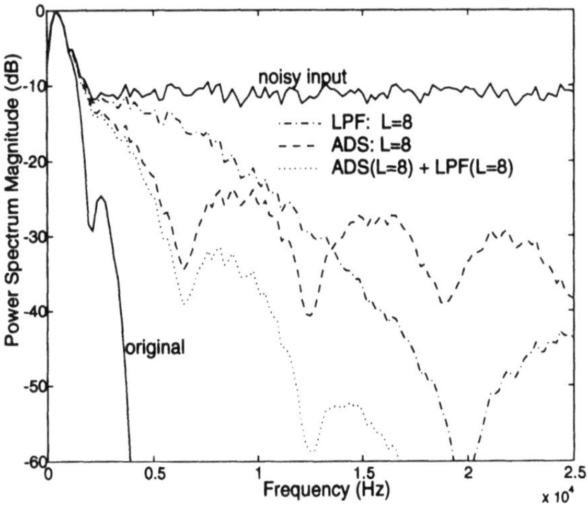


(a)

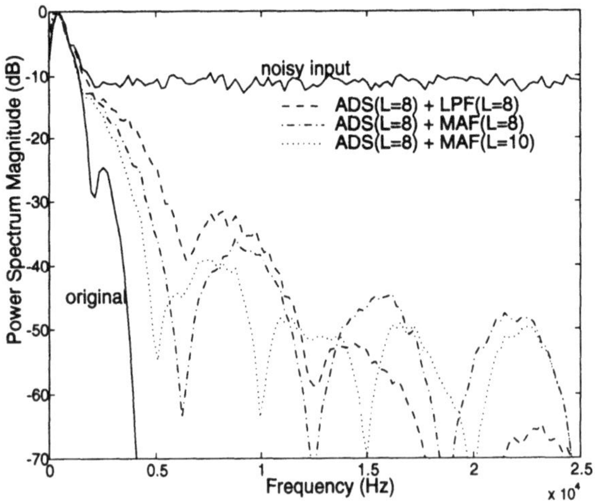


(b)

**Figure 5.12** *(a) ADS vs. LPF. (b) ADS + LPF/MAF.*

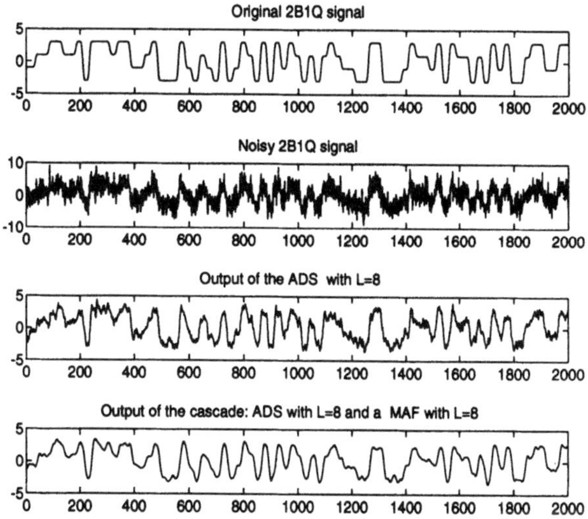


(a)

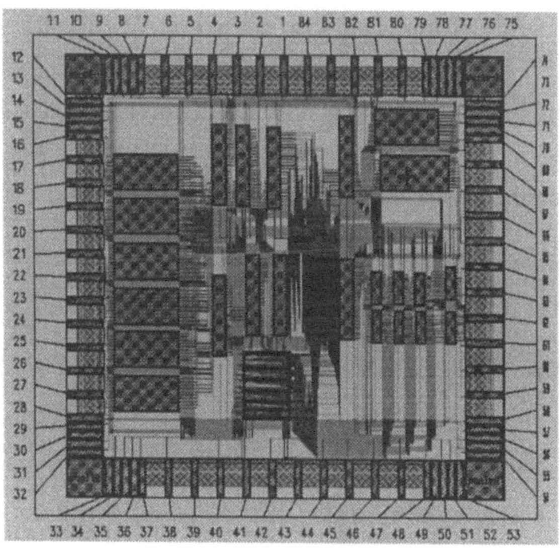

(b)

**Figure 5.13** *(a) ADS performance, (b) The 0.8μ 50 Mhz ADS chip.*

In this section, we describe the rapid prototyping of FFT processors within the core-based design environment proposed earlier. Several single-chip real-time FFT processors have been designed in very short design times (a few weeks). A high performance FFT macro-component, COMAS (library hierarchy LEVEL-2), is used as a core for the the datapath of the proposed FFT processors. Figure 4.9 illustrates the design approach to prototyping FFT processor within the proposed environment. The FFT core, COMAS, is a decimation-in-time radix-2 butterfly module, as illustrated by Figure 5.14.

### 5.2.1 Architecture

The proposed FFT core, *a quad-butterfly architecture* (library hierarchy LEVEL-2), is illustrated by Figure 5.14. This core was inspired from the FFT work done in [91]. The quad-butterfly architecture combines maximum parallelism, pipelining, and butterfly cascading. While memory access seems to be the major bottleneck to achieving real-time performance, the core described here has an efficient memory access pattern.

As illustrated by Figure 5.14, the quad-butterfly architecture is composed of:

1. Quad-Butterfly Execution Unit (QBEU),
2. IN/OUT data RAMs, coefficient ROMs,
3. Control Unit (CU),
4. Data Address Generators (DAG), and Coefficient Address Generators (CAG).

#### *Execution unit*

The execution unit (QBEU) of the quad-butterfly architecture of Figure 5.14 is composed of four butterfly computational units or COMAS cores. Each COMAS core (see figure 5.14) implements a decimation-in-time radix-2 butterfly structure. The complex-valued outputs, $A'$ and $B'$, of this butterfly structure are computed from the complex-valued inputs, $A$ and $B$, as follows:

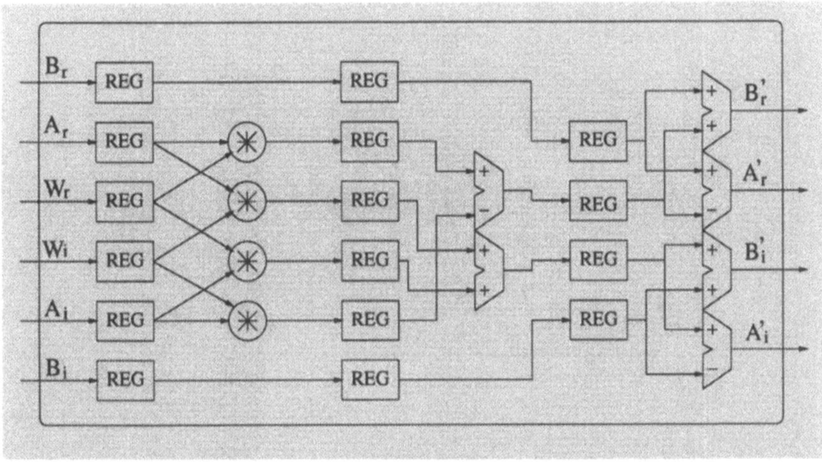


**Figure 5.14** *COMAS: a decimation-in-time radix-2 butterfly core (library hierarchy LEVEL-1).*

$$B' = B_r' + jB_i' = B_r + A_r W_r - A_i W_i + j(B_i + A_r W_i + A_i W_r), \qquad (5.9)$$

$$A' = A_r' + jA_i' = B_r - A_r W_r + A_i W_i + j(B_i - (A_r W_i + A_i W_r). \qquad (5.10)$$

Here $W = W_r + jW_i$ is the twiddle factor.

The COMAS cores that implement the QBEU are cascaded such that saving into the output RAMs is needed only every other FFT stage. The computational path of this structure is illustrated by Figure 5.15(a).

Four complex butterflies (two from one stage and two from the subsequent stage) are executed during one clock cycle. A vertical execution path is followed by this structure. That is, when the execution of two consecutive FFT stages are completed the QBEU proceeds with the computation of the two following stages. Every computational node in the radix-2 butterfly structure is mapped to one dedicated hardware core. Registers are added in between the computational nodes for pipelining purposes. Hence, a COMAS module is composed of four multipliers and six adders and subtracters. The QBEU is

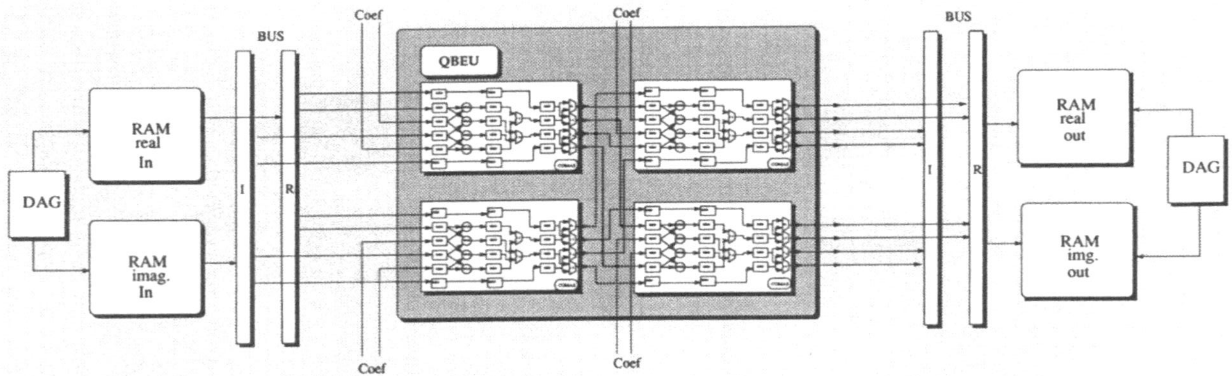


Figure 5.12: *The quad-butterfly architecture: input RAMs, output RAMs, and the core of the architecture is a datapath that consists of 4 butterflies (QBEU) (hierarchy LEVEL-2).*

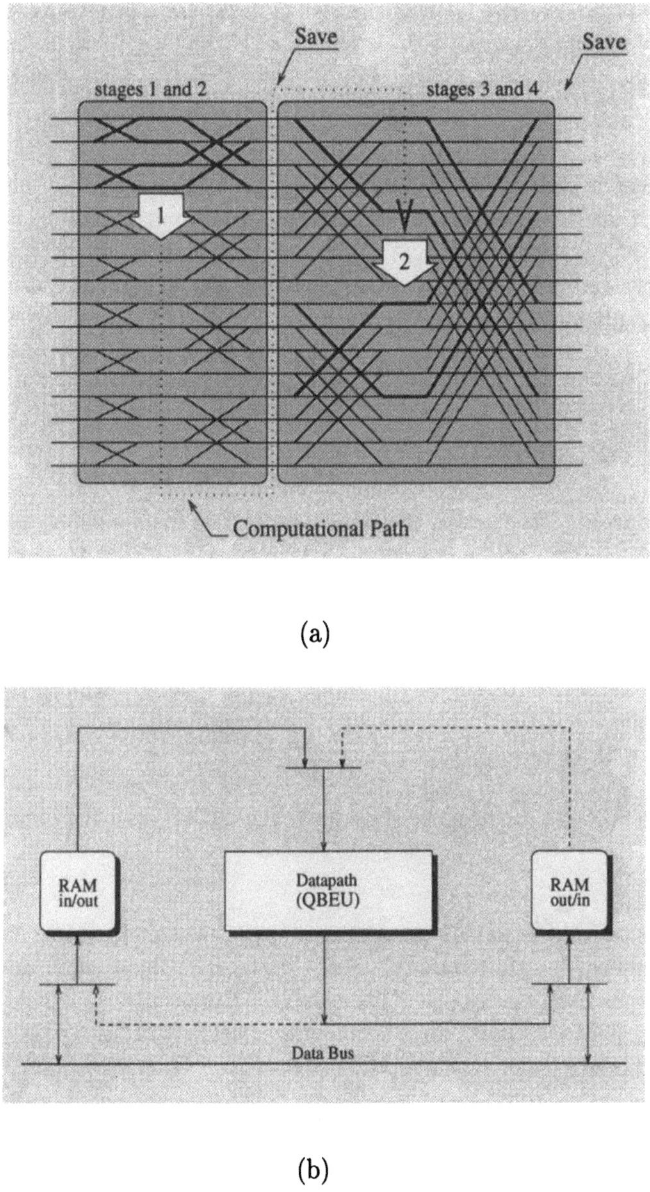


**Figure 5.15** *(a) Two FFT stages, (b) Memory configuration pattern.*

composed of four COMAS modules[2]. Maximum pipelining is realized by inserting registers in between all successive computational nodes, such that the critical path ($I_{cycle}$) of the QBEU is matched with the computational delay of one single multiplier ($d_{mult}$):

$$I_{cycle} = d_{mult} \tag{5.11}$$

The QBEU described above executes 40 fixed-point operations per clock cycle. Considering that the array multiplier used has a computational delay of $40ns$, the QBEU delivers 1000MOPS. A superior performance can be obtained by adopting intra-cell pipelining[3], or by using higher performance functional units (i.e., booth multiplier).

## *Memory bandwidth*

The highly parallel structure of the QBEU requires a wide memory bandwidth. Eight operand fetches and eight operand writes need to execute at every clock cycle. Separate memory spaces are allocated for real and imaginary data, and also for read (input RAMs) and write (output RAMs) operations. The configuration of the input/output RAMs is illustrated by Figure 5.15(b). After every two computational FFT stages, the input/output configuration of the RAMs is reversed and the QBEU proceeds with the computation of the two following FFT stages with the new in/out configuration.

Different alternatives are proposed to supply the QBEU with the required memory bandwidth:

- 4-ported RAMs — in this case the read/write cycle is the same as the computational cycle. The input data RAMs (real & imaginary) are accessed once during one clock cycle. The same for the output data RAMs. Eight data reads (4 real parts and 4 imaginary parts) and eight data writes are executed during one clock cycle.
- 2-ported RAMs — Here the read/write cycle is half the computational cycle. That is, RAMs are accessed twice during one clock cycle for read and write operations.

---

[2]a total of 16 multipliers and 24 adders and subtracters are used

[3]pipelining inside the functional units

- 1-ported RAMs — In this case the RAMs are accessed 4 times during each clock cycle for read and write operations.

For instance, the first version (FFT-V1.0) of the FFT chip series designed within the the environment adopts the solution with 1-ported RAMs. A high dynamic range is supported by adopting 24-bit fixed point data type with one sign bit and two guard bits to avoid overflow. Four 1024-Word X 24-bit 1-ported RAMs are used for storage. Two are configured as input RAMs (real & imaginary), and two are configured as output RAMs (real & imaginary). The access time of the adopted 1-ported RAMs is estimated $15ns$. Since every RAM is accessed 4 times during one clock cycle, a clock cycle of at least $60ns$ is required. FFT-V1.0 has a clock cycle of $I_{cycle} = 80ns$. Considering this clock cycle the QBEU is working only at 50% of its maximum speed. That is, it delivers only 500MOPS[4].

## *Data address generators*

The flow diagram of the data address generator (DAG) for the proposed architecture is illustrated in Figure 5.16. Registers $R1$, $R2$, $R3$, *and* $R4$, hold a new operand address at every clock cycle. Real and imaginary data are accessed at the same addresses but in different memories. A snapshot of the simulation results of the DAG block is illustrated in Figure 5.19(b). A coefficient address generator (CAG) is also used to generate the twiddle factor addresses such that four coefficients are read from the ROMs every clock cycle. The flow diagram model for the CAG is similar to the DAG model of Figure 5.16. The coefficients are stored in ROMs. Real and Imaginary parts of the coefficients are stored in different memory spaces and are accessed by the same addresses. FFT-V1.0 uses two 1024-Word X 24-bit ROMs (real & imaginary).

## *Controller*

The VHDL core for the control is illustrated by Figure 5.17. Three different processes are used:

[4] in this case the same performance can be obtained with only 2 COMAS cores

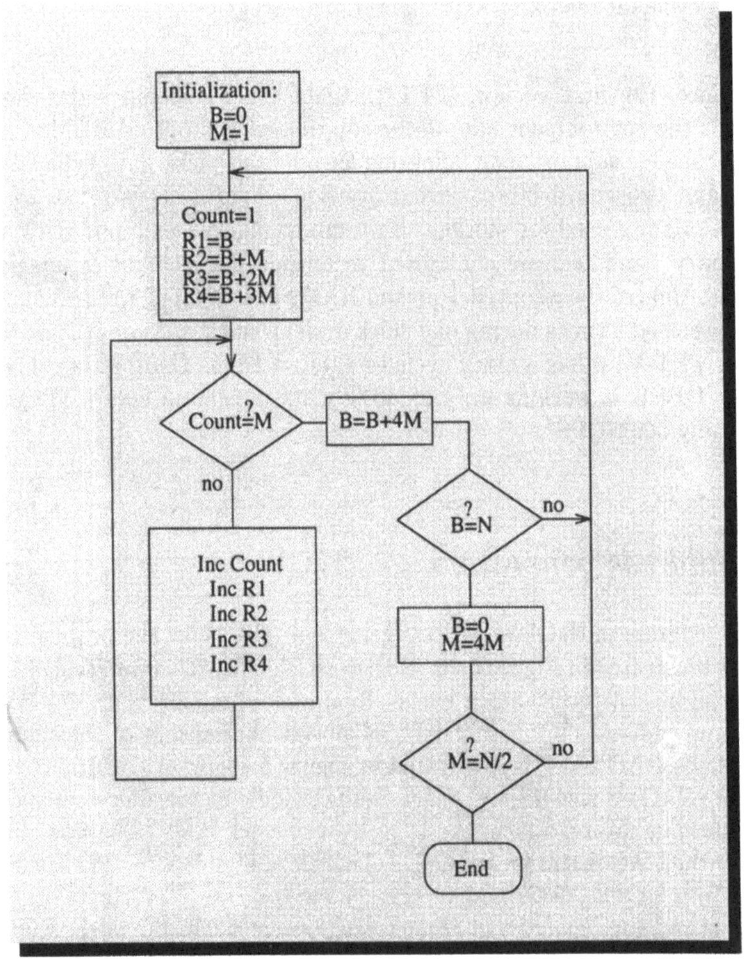


**Figure 5.16** *Flow diagram of the Data Address Generator (DAG) of the FFT chip.*

- *Next_STATE*– is a combinatorial process that is sensitive to the *CURRENTSTATE* signal and to all the inputs. it is implemented with a *case* statement that percolates through all the defined states of the FSM, and

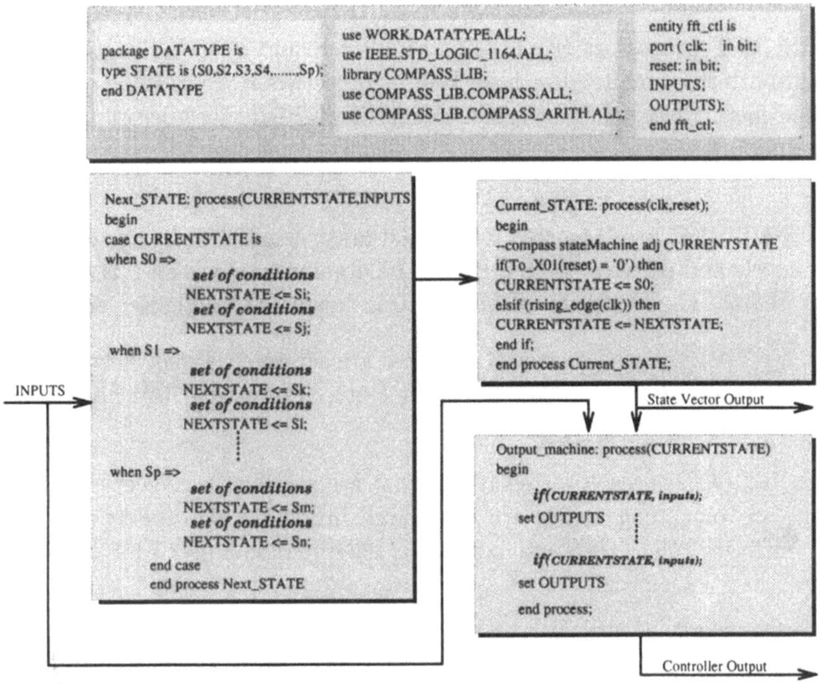


**Figure 5.17** *A VHDL model for the Controller of the FFT chip.*

assigns a specific state to the $NEXTSTATE$ signal based on the current state and the inputs.

- *Current_STATE*– is a sequential process that, at every clock cycle, assigns the $NEXTSTATE$ signal to the $CURRENTSTATE$ signal. This process also allows for an asynchronous reset of the FSM through the *reset* signal.

- *Output_machine*– is the process that generates the control signal of the FSM based on the $CURRENTSTATE$ signal and the inputs to the machine.

Figure 5.19(a) illustrates the simulation results of this FSM.

### 5.2.2 Applications

Three FFT processors (FFT-V1.0, FFT-V1.1, and FFT-V1.2) based on the quad-butterfly architecture have been synthesized in COMPASS using core-based methodologies. They differ mainly in their memory access pattern and pipelining structure:

- FFT-V1.0: uses 1-ported RAMs that are accessed 4 times during one clock cycle and delivers $500MOPS$. A 1024-point complex FFT is executed in $108\mu s$. The Floorplan of FFT-V1.0 is illustrated by Figure 5.20(b).
- FFT-V1.1: uses 2-ported RAMs that are accessed 2 times during one clock cycle and delivers $1000MOPS$. FFT-V1.1 executes a 1024-point complex FFT in $52\mu s$.
- FFT-V1.2: uses 4-ported RAMs that are accessed once every clock cycle. Two pipelining stages are introduced inside the multipliers. FFT-V1.2 delivers *2000MOPS* and executes a 1024-point complex FFT in $27\mu s$.

The performance of the proposed chips is inversely affected by the pipelining overhead as the FFT size decreases. The effective performance of FFT-V1.0 and FFT-V1.2 as function of the FFT size is illustrated by Figure 5.18(a). For the FFT-V1.0 version, the pipelining overhead is about 5% for a 1024-point FFT and attains 75% for a 64-point FFT. In the case of the FFT-V1.2 version, "shadow" RAMs are used to handle the switching of the configuration of the in/out RAMs without causing any interruption in the pipeline. This added feature results in an improvement in the pipelining overhead which becomes only 0.8% for a 1024-FFT and about 21% for a 64-point FFT. Performance comparison of the synthesized FFT chips with state-of-the-art single-chip dedicated FFT processors (such as O'Brien's [89], NTT's [90]) is provided by Tables 5.18.

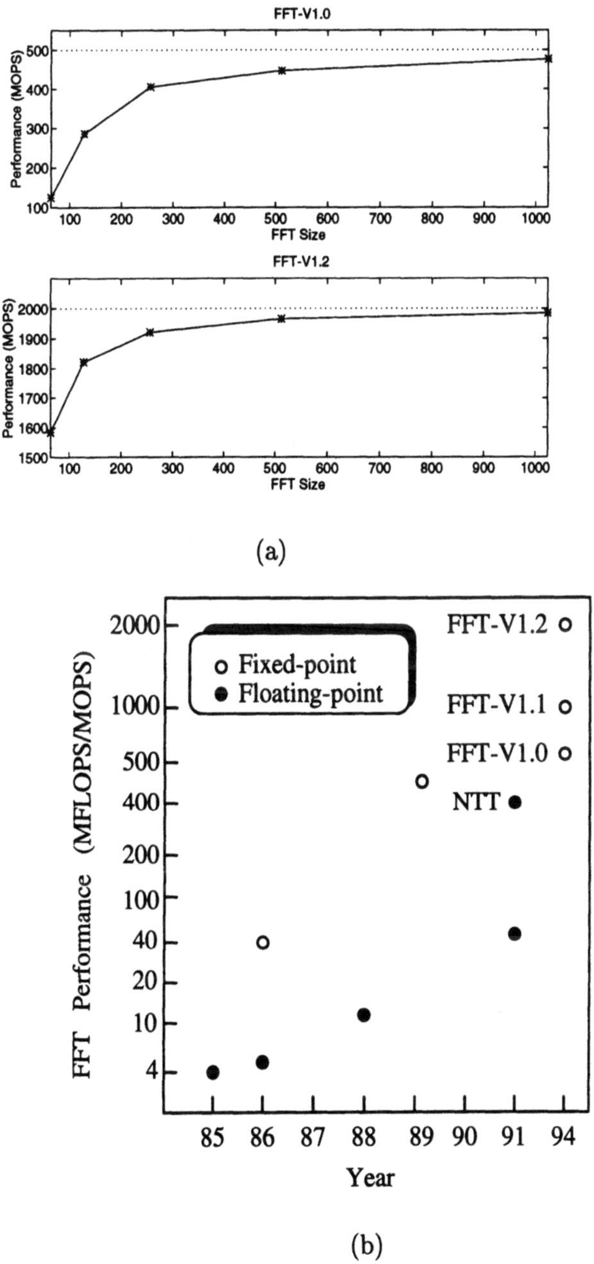


**Figure 5.18** *(a) FFT-V1.0 and FFT-V1.2 chipset, (b) Comparison with commercial FFT chips.*

Table 5.1: Performance comparison with state-of-the art FFT Processors.

| Chip | CMOS | Design | Clock | Area $mm^2$ | Data format | Performance |
|---|---|---|---|---|---|---|
| O'Brien's | $1.4\mu$ | schematic | 40MHz | 13.16x13.66‡ | 16-bit fixed-pt | 200 MIPS |
| NTT's | $0.8\mu$ | schematic | 40MHz | 11.58x11.58§ | 24-bit float-pt | 400 MFLOPS |
| FFT-V1.0 | $0.8\mu$ | VHDL | 12.5MHz | 10.87x10.87‡ | 24-bit fixed-pt | 500 MOPS |
| FFT-V1.1 | $0.6\mu$ | VHDL | 25MHz | - | 24-bit fixed-pt | 1000 MOPS |
| FFT-V1.2 | $0.6\mu$ | VHDL | 50MHz | - | 24-bit fixed-pt | 2000 MOPS |

‡core size, §chip size

Table 5.2: Execution time comparison of a 1024-point complex FFT.

| TMS320C40 | DEC alpha | ADSP21060 | O'Brien's† | FFT-V1.0 | FFT-V1.1 | FFT-V1.2 |
|---|---|---|---|---|---|---|
| $1545\mu s$ | $500\mu s$ | $460\mu s$ | $100\mu s$ | $108\mu s$ | $52\mu s$ | $27\mu s$ |

†radix 4

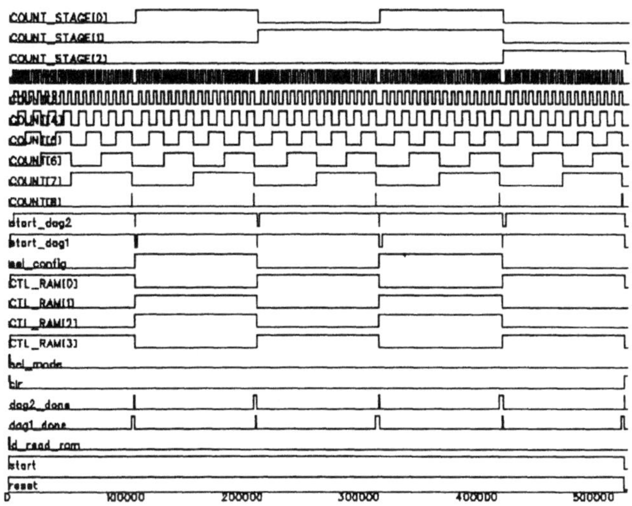


(a)

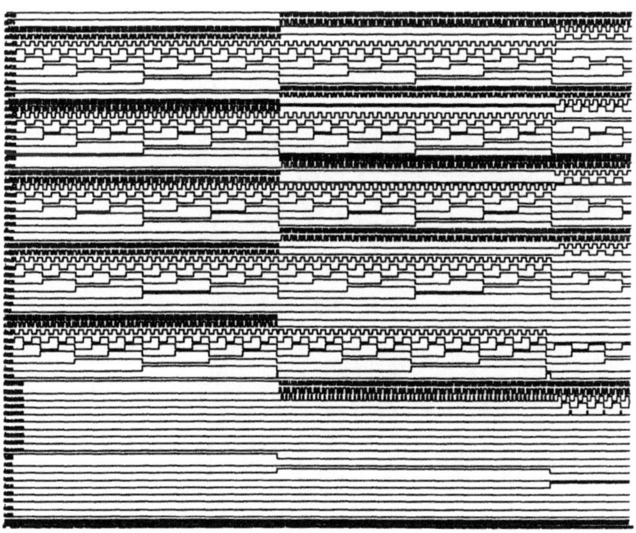

(b)

**Figure 5.19** *(a) Simulation of control unit (CU). (b) Simulation of data address generator (DAG).*

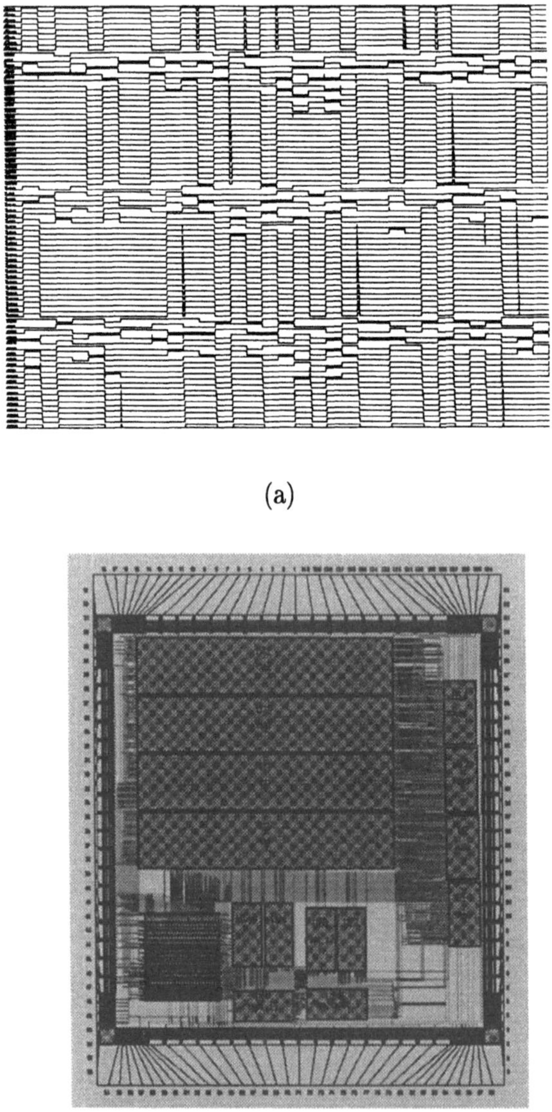

(a)

(b)

**Figure 5.20** *(a) Simulation of QBEU core. (b)* 12.5*Mhz FFT (FFT-V1.0).*

# 6

# BOARD INTEGRATION

Most ASICs fail within the board-level environment if the component interface clock-level timing and functionality is not considered early in the design process. Recent AT&T studies show that nearly 50% of ASICs fail at the system level. This chapter describes how core-based ASICs can be integrated into board-level designs.

## 6.1 INTRODUCTION

A board or an MCM packages microprocessors, memories, controllers, ASICs, busses, glue logic, and interfaces to the external environment. The design of a typical board usually spans over a design period of 2 – 3 months, at the cost of 0.5 person-year [93]. Board integration can also be viewed as a hardware/software integration task that has to support the concurrent and efficient development of hardware and software in order to achieve a desired functionality and performance goals [97]. New design methodologies, such as hardware-less Virtual Prototyping [92], have been proposed as part of ARPA's RASSP program as a means to achieve efficient and timely board and rack level HW/SW integration and test.

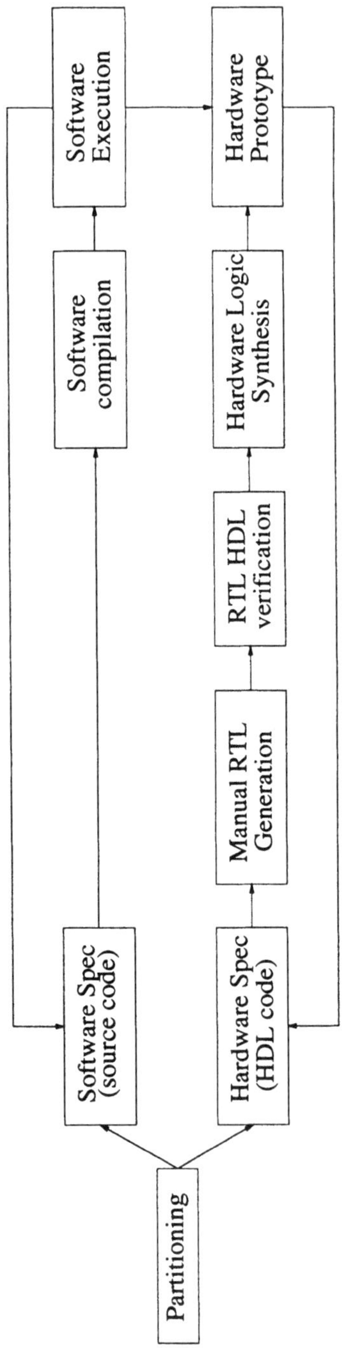


Figure 6.1: *Current practice in hardware/software co-design is characterized by a sequential bottleneck--software design waits on the development of a hardware prototype.*

## 6.2 CURRENT DESIGN PRACTICE

Current practice in board integration is characterized by very little interaction between the hardware and software design cycles. Board integration does not start before design and fabrication of the hardware is completed. Software testing is typically done on emulator boards that model the components of the target system. Late integration of software and hardware elements causes a number of problems that include — (1) software makes incorrect assumptions about hardware implementation, (2) software is less maintainable due to corrections and patches to account for hardware errors at integration time, and, (3) the ASIC may have to be re-designed during the integration phase, causing long schedule delays and cost overruns.

An illustration of current practice in board integration is given by Figure 6.2 [97]. The cyclic coupling between hardware and software designs greatly penalizes integration and system test.

## 6.3 VIRTUAL PROTOTYPING

ARPA's RASSP program provided many insights into the methodology of *Virtual Prototyping* [92]. Virtual prototyping synthesizes a clock cycle-accurate VHDL model of the target system prior to fabrication. Software is then executed on the virtual system to ensure that errors in design are corrected before fabrication is started. A candidate RASSP design methodology proposed by Madisetti and Egolf is illustrated in Figure 6.1 [93] that defines a systematic approach to efficient board level design.

The design efficiency of virtual prototyping is a direct reflection on the accuracy of the provided VHDL models for ASIC cores and processor elements.

## 6.4 BOARD-LEVEL MODELING

VHDL modeling for design can be performed at many levels of abstraction. A survey of approaches to modeling HW/SW systems within a virtual prototyping

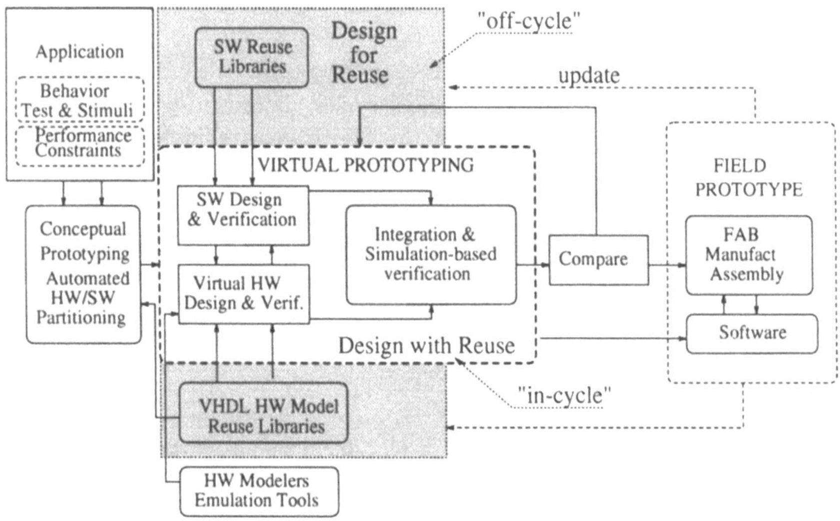


**Figure 6.1** [93] *System design based on virtual prototyping and extensive reuse.*

environment is available in [96]. Figure 6.2 [96] illustrates the different phases in the top-down virtual prototyping process. The first step is the creation of a *functional* virtual prototype that represents an executable specification in VHDL. The model describes the functional behavior and the test bench.

Peformance-level virtual prototyping investigates the efficiency of mapping of the task graph to the embedded system architecture through simulation. Performance modeling uses abstract timing libraries similar to those defined in [98]. Performance models are not synthesizable, and model coarse timing of various components (usually in units of ns). For instance pipelines are modeled as a delays, memories are modeled as with an access time delay, bus interface units are characterized by a data transmission rate, I/O devices are characterized by a transmission rate, and processors are defined through their task scheduler and set of resources [96]. Processor, bus utilization, throughput, and memory usage are performance measures evaluated at the performance level.

The next and more detailed step in the virtual prototyping process is fully functional modeling. The functionality and clock-level timing of the SW/HW

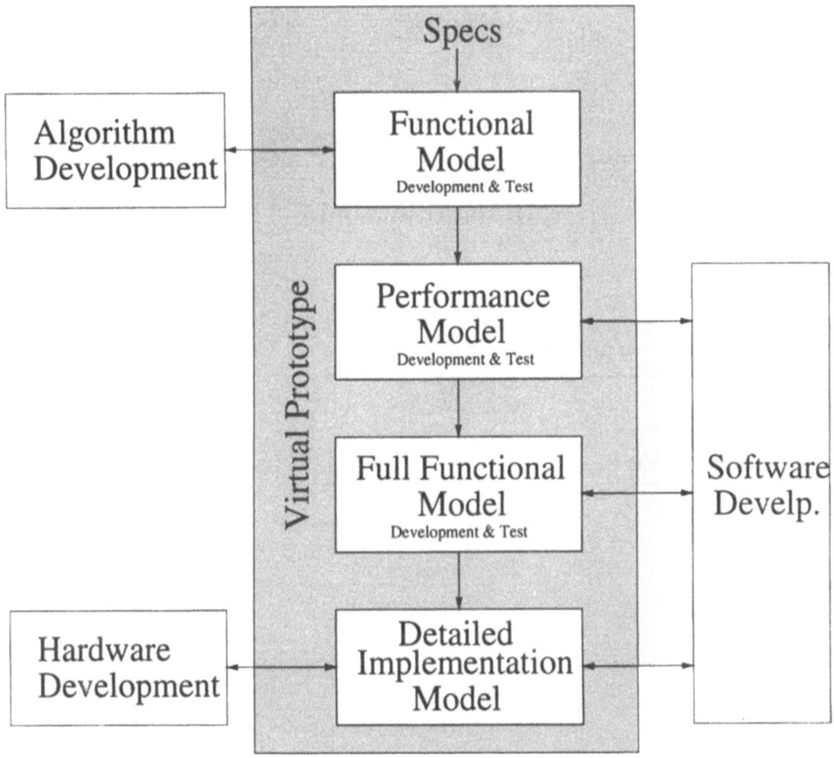


**Figure 6.2** *Phases in the virtual prototyping process.*

system is modeled, and it utilized to perform detailed hardware and software design tasks. System integration and test is greatly streamlined at this step.

The final step of the virtual prototype modeling is to convert the parts of the design that are meant to be implemented in hardware into a synthesizable version. This final stage of the virtual prototyping is the detailed implementation level.

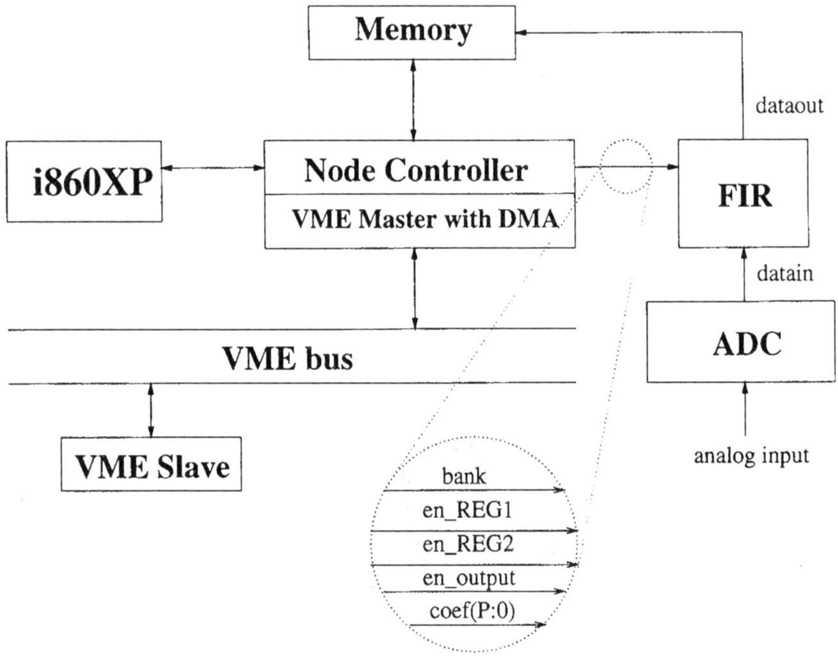


**Figure 6.3** *Block-diagram of an i860-based board containing FASICs.*

## 6.5 A CASE STUDY

The board-level integration of the core-based ASICs designed within the proposed environment starts at an early stage of the system integration cycle. There is no need to wait for fabrication of the designed ASICs. The VHDL model of the ASIC cores are used instead as will be described in case study that follows.

A case-study of an i860-based board level integration is illustrated by Figure 6.3. A FASIC (FIR filter-RTL model) designed within the proposed environment is interfaced to an all-VHDL behavioral board model. The board is based on the i860XP processor model described in [92]. The FASIC is a FIR filter that is used as a coprocessor. The board is composed of a variety of behavioral VHDL models developed at Georgia Tech — i860XP processor (behavioral VHDL) model, a memory model, a node controller model, and a VME bus model. The internal model of the i860XP provides for instruction fetch, decode, execution and write back [94]. The node controller interprets the data (or instruction transfer type) for all processor operations that transfer to the bus. When the i860XP needs to initiate a filtering operation on the FASIC chip, it sends a *chip_select* signal to the FIR through the node controller. The FIR filter receives its incoming data from an Analog-to-Digital Converter (ADC), processes it, and loads it back to the memory. When the FIR filter is done with the filtering task, it interrupts the i860XP processor that decides when to access the filtered data. The board-level system was debugged completely in about 4 person-hours. The i860-based board and the FIR model were designed and fully simulated prior to the board integration. A snapshot of the system-level simulation is illustrated by Figure 6.6 showing the details in functionality and clock level timing long before any fabrication is done.

Every element in the board is described as a *component* that interconnects with the other components of the board. The components and signals in the board level virtual prototype include:

- an *i860 component* that models the internal structure of the i860 processor.
- a *CLK_RESET_GEN component* for clock and reset signal generation.
- a *MEMORY component* that holds the application program and other useful information. A user written binary file that describes the application is automatically loaded into the memory component at the *reset* signal.
- a *NODE_CONTROLLER component* interfaces between the i860 processor and the rest of the components (nodes) of the system. It interprets the instructions of the i860 processor and sends the appropriate control signals and data information to the node.
- a *VME_MASTER_SLAVE component* mediates the interactions between master and slave via the VME bus.
- an ARBITER component resolves contention. It selects between the master nodes that request access to the VME bus at the same time.
- a *REQUESTER component* file requests master nodes to access slave nodes. Requests are filed to the *ARBITER component.*
- a *BUSTIMER component* measures the time for data transactions through the VME bus. Timeout errors are generated when transaction times last longer than a default threshold.

```
--i860_board.vhdl
--library declarations
entity i860_board is
end i860_board;
architecture STRUCTURAL  of i860_board is
      component ADC
            generic (half_clk_period : TIME := 400 ns);
            port (dataout : out bit_vector(P downto 0);
                  clk2     : out bit);
      end component;
      for all: data_src use entity SYSTEM.data_src(behave);
      component cell1
            port (datain        : in bit_vector(P downto 0);
                  clk           : in bit;
                  bank          : in bit;
                  en_REG1       : in bit;
                  en_REG2       : in bit;
                  en_output     : in bit;
                  coef          : in bit_vector(P downto 0);
                  acc           : in bit_vector(M downto 0) ;
                  dataout       : out bit_vector(M downto 0));
       end component;
      for all: cell1 use entity SYSTEM.cell1(behave);
      component NODE_CONTROILER
            port (bank          : out bit;
                  en_REG1       : out bit;
                  en_REG2       : out bit;
                  en_output     : out bit;
                  coef          : out bit_vector(P downto 0));
      end component;
      for all:  NODE_CONTROLLER
      use entity i860XP.NODE_CONTROLLER(behavior);
  ---Declarationof the rest of the components:
      ::::::::::::::::::::::::::::::
      begin
      -----PORT MAP CONNECTIONS-----
      ::::::::::::::::::::::::::::::
      end STRUCTURAL;
```

**Figure 6.4** *Virtual prototyping: case study.*

```
--NODE_CONTROLLER.vhdl
--library declarations
use i860XP.passpara.all;
entity NODE_CONTROLLER is
--generic declaration
port (--other port declaration
        --FIR filter interface
       bank     : out bit;
       en_REG1 : out bit;
       en_REG2 : out bit;
       en_output : out bit;
       coef        : out bit_vector(P downto o));
end NODE_CONTROLLER:
architecture behavior of NODE_CONTROLLER is
::::::::::::
VME2: process
    begin
       if (REST = '1') then
       --initialization
       end if;
       if(ADS_N'EVENT and ADS_N = '0') then
       TEMP_ADDR : std2bit_vector("000" & ADDRESS, '0');
       if (TEMP_ADDR = X"00000001" or --config register 1
       TEMP_ADDR = X"0000000 F" or -- FIR configuration
       --other configurations ) then  --Address register
       wait until ADS_N'EVENT and ADS_N= '1';
       ::::::::::
    --FIR setup    --Bits P:0 for the coefficient
    coef <= std2bit_vector (DATA(P downto 0),'0');
    bank <= std2bit(DATA(P+1),'0');--Bit P+1 is the bank
    en_REG1<= std2bit(DATA(P+2),'0');--Bit P+2 is en_REG1
    en_REG2 <= std2bit(DATA(P+3),'0');--Bit P+3 is en_REG2
    en_output<= std2bit(DATA(P+4),'0');--Bit P+4 is en_output
    :::::::::::
end process VME2;
end behavior;
```

**Figure 6.5** *The node controller.*

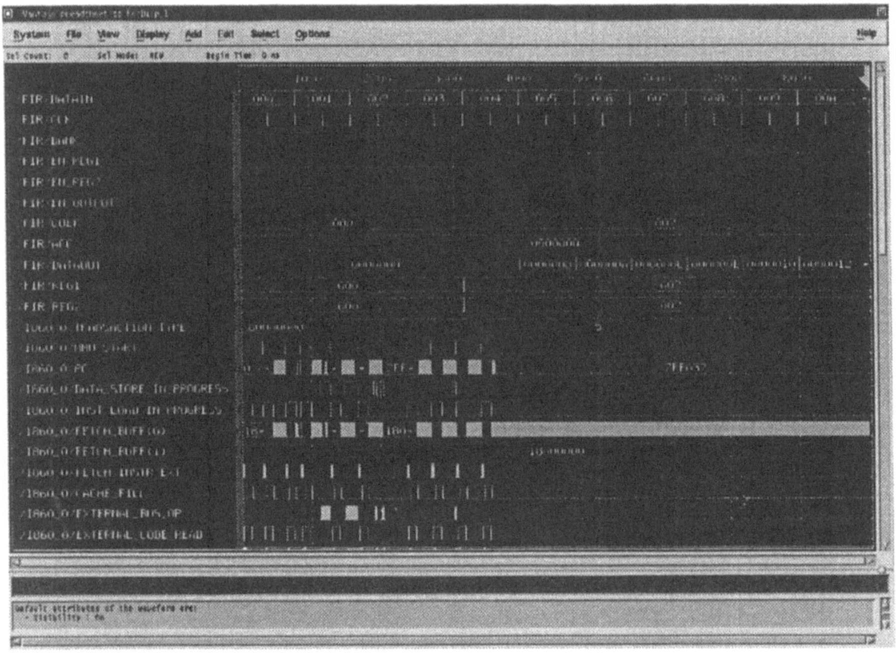

**Figure 6.6** *Board level simulation with the i860 and FASICs. The master processor loads the coefficients into the FIR filter, then orders the filtering process. The functionality of the FIR filter within the system is verified. The output of the filter, DATAOUT, is the result of the filtering of the input, DATAIN.*

To this board level virtual prototype a FASIC FIR that was designed in earlier chapters was added. Two new components are added to the board-level virtual prototype:

- an *ADC component* drives the data to be filtered by the FIR filter. This component emulates an analog-to-digital converter.

- a *cell1 component* filters the incoming data at the request the i860 master processor.

Hardware integration consists of making the appropriate hardware interconnections with the rest of the board. The new elements are tied to the system through their entities. Figure 6.3 illustrates how the new FIR filter (*cell*1) is integrated into the system. The *Node controller* takes charge of interfacing the new component to the master processor. This involves a hardware and software extension to the existent *Node controller*. The hardware extension results in new output pins to drive the FIR filter through its entity. The software extension consists of adding a new addressable node to the existing protocol of the *Node controller*. The output of the filter is connected to the existent memory. The access to the memory for *write* operations is also regulated by the *node controller*. Excerpts form the board and the Node controller are illustrated by Figures 6.4 and 6.5, respectively.

The software integration consists of programming the i860XP processor to take advantage of its new co-processor. At first, no cache hits are registered for the i860XP processor. The address for the first instruction is issued for the memory through the *Node controller*. The program is written by the system designer and stored in a binary format. At the reset signal, the binary file is automatically loaded into the memory component.

## 6.6 CONCLUSION

Board integration via virtual prototyping has proved its effectiveness through several success stories [94, 96, 92]. All-VHDL virtual prototyping has bridged the design gap that has always existed between hardware and software, resulting in short design cycles, portability, easy upgrading and integration. FASICs are easily integrated into the system, and their performance and functionality are validated in the system before hardware prototypes are attempted.

# 7 CONCLUSIONS

Application Specific Integrated Circuits (ASICs) have kept pace pace with the increasing complexity of applications and their demands for higher data rates, though this situation seems ready to change. Estimates show that the capability (in terms of gates) to design efficiently would lag considerably behind the capability to manufacture ASICs. New factors that influence the design of ASICs are also to be taken into consideration. These include; deep submicron designs, developments in synthesis and top-down design using hardware description languages (HDLs), and increasing time-to-market pressures. The primary goal is to provide users with desirable features such as — flexibility, ease-of-use, high reuse, quick-turnaround, high performance, integrability, accuracy, portability, and reliability. Existing design methodologies and environments support a few of these desirable features (see section 2.3). The purpose of this monograph was to define a design methodology for ASSPs that provides a competitive advantage to an organization through the use of commercial-off-the shelf (COTS) CAD/ESDA design environments.

## 7.1 HIGHLIGHTS

The highlights of the approach proposed in this monograph are listed below.

### 7.1.1 Core-based synthesis of ASSPs

A sytematic design methodology for ASICs that utilizes organization-specific and application-specific cores and reusable cores has been proposed and validated. Efficient functional reusability is derived from the built-in flexibility of parameterized design with VHDL [15], and its many powerful constructs and features (i.e., abstraction, generalization, encapsulation, inheritance, and reuse). Efficient and rapid design exploration is realized through a limited but realistic design space exploration (and defined by the organization or customer) within a reusable datapath and controller library. The exploration of the design space is implemented via a tree-based search algorithm, and other algorithms bear investigation. Once an application class is chosen, the design alternatives are listed horizontally, and a variation of implementations of every alternative are pursued vertically. Design alternatives represent a variation of architectures (sequential, parallel). Each design alternative spans vertically to a variety of implementation issues (i.e., pipelining, multiplexing). Reuse is maximized so that at the top of the tree a design architecture is linked to pre-designed, pre-verified domain-specific macro components (i.e., FFT architecture calls butterfly core components). These cores themselves call upon datapath sub-components of smaller sizes. The parameters that instantiate the design spread vertically into the tree.

Reuse of design entities at the functional level facilitates a higher payoff in terms of design time, ease-of-use, and performance. Productivity is greatly enhanced when the designer can work with a number of reusable cores and entities that are optimized for the customer's application and goals. Teams of designers are quickly able to leverage off each others design efforts, leading to enhanced results and quality of designs. Much less time is also spent on the validation of the functionality of a design that is built with pre-design pre-tested building blocks.

### 7.1.2 The FASIC design environment

The FASIC approach of this monograph has been implemented in VHDL, and a graphical user interface has been provided to enhance its ease-of-use. The proposed environment considers three design views: (1) The *core or macrocell* design view defines the Design-for-Reuse (DfR) cycle. Guidelines and rules are set to build a reusable library that provides ease-of-use, flexibility, and efficiency. The library environment proposed is based on a hierarchical structure of several levels. (2) The *system* design view defines the Design-with-Reuse (DwR) cycle. Component and core identification, classification, storage, and retrieval are handled systematically. A Tcl/Tk graphical user interface is offered for this purpose. The designer is invited to use one of two design lines: (a) the *automatic design line* (ADL) is a collection of turn-key tools for deterministic DSP algorithms (i.e., FFT, FIR, Viterbi, etc.). One can explore design alternatives through a tree-based search algorithm of a variety of reusable architectures and cores. (b) the *functional design line* (FDL) is an environment where applications are designed at the multiple VHDL entity-level using a functional decomposition of the application. The defined entities are then mapped to the corresponding functional blocks or cores from a reusable library. (3) The *simulation* design view relates to the verification of the functionality and performance to the original specifications.

The proposed approach promises the reduction of the design turnaround cycle of ASICs and ASSPs by a factor ranging from "4X" to "100X" depending on how much customization is done in-cycle.

### 7.1.3 A DfR library of cores

A DfR library of cores has been designed to support the proposed FASIC approach. The library is based on a hierarchical structure of several levels. The lowest hierarchy level is composed of elementary datapath components (i.e., adders, registers, multipliers). At the next hierarchy level, the functional level, domain-specific datapath cores are defined as building blocks for DSP-type applications. These cores could utilize the lower level datapath components, however, for deep submicron designs, could represent custom-designed (off-cycle) building blocks and cores. At a higher level of abstraction multiple entities/architectures in VHDL are provided. These reusable architectures use the cores in the DfR library. Every component in the library is characterized by its domain, its structure, and its features. The domain defines the algorithm (i.e., FIR, FFT), the structure defines the architecture (i.e., sequential, parallel), and the features state the variations of the reusable block for fixed domain and structure (i.e., degree of multiplexing, level of pipelining). The proposed core library encompasses the design of non-adaptive and adaptive (LMS and Block-LMS) digital filters and FFT processors.

Cores should be flexibly parameterized so as to capture a range of application specifications that are of interest to the organization. For example, a variation of the degree of time multiplexing can be utilized optimize the area, while increasing the level of pipelining (inter-, and intra-primitive cells) allows for high data rates. Power dissipation can be optimized by reducing the supply voltage (i.e., the throughput is kept constant via combining pipelining with a greater circuit parallelism), tagging the design to a low power primitive cell library, or choosing from a range of multiple parallel functional cores.

Each core in the proposed reuse library is annotated with information about the expected area, speed, and power for different technologies and architectural styles. Estimation of design cost information about applications that use components from the library is thus streamlined, and currently a fuzzy set-based front end tool is being developed at Georgia Tech to assist with design exploration based on customer specifications.

### 7.1.4 Business models

A business model was utilized to highlight the advantages and the limitations of the proposed core-based ASIC synthesis approach. The cost model is based on the derivation of a break-even analysis that is quantified in terms of product volume over which one design alternative is considered more profitable than the other.

### 7.1.5 Validation of approach

Using the design environment proposed, several high performance DSP-ASICs have been designed and tested in design turnaround periods varying from one day to a few weeks. These were discussed in previous chapters and include:

- A 60-tap Block-LMS FIR filter was designed in $0.8\mu$ CMOS technology, and tested for a maximum clock frequency of $50MHz$, and a maximum (video) data rate of $12.5Mbaud$, with a core area of $81.95mm^2$, and a power dissipation estimated to $1.387W$.
- A 64-tap programmable FIR filter was designed in $0.8\mu$ CMOS technology, and tested for a maximum operation frequency of $50MHz$, and a maxi-

mum data rate of $12.5Mbaud$, with a core area of $61.35mm^2$, and a power dissipation estimated to $1.331W$.

- A 5-stage IIR lattice filter (time scaled by two) has been designed in $0.8\mu$ CMOS technology. The filter occupies $15.20mm^2$ of silicon area, and dissipates $0.25W$ of power at an operation frequency of $25MHz$.
- A 8-tap All Digital Sensor (ADS) chip was designed in $0.8\mu$ CMOS technology. The ADS chip implements the ADS algorithm described in [83], occupies an area of $16.51mm^2$, and operates at a frequency of $50MHz$.
- Three FFT processors (FFT-V1.0, FFT-V1.1, and FFT-V1.2) based on the quad-butterfly architecture have been designed within the proposed environment. They differ mainly in their memory access pattern and pipelining structure:
    1. FFT-V1.0: uses 1-ported RAMs that are accessed 4 times during one clock cycle and delivers $500MOPS$. A 1024-point complex FFT is executed in $108\mu s$. The silicon area and the frequency of operation of that chip are 118.183 $mm^2$ and $12.5MHz$, respectively.
    2. FFT-V1.1: uses 2-ported RAMs that are accessed 2 times during one clock cycle and delivers $1000MOPS$. FFT-V1.1 executes a 1024-point complex FFT in $52\mu s$.
    3. FFT-V1.2: uses 4-ported RAMs that are accessed once every clock cycle. Two pipelining stages are introduced inside the multipliers. FFT-V1.2 delivers $2000MOPS$ and executes a 1024-point complex FFT in $27\mu s$.

The chips designed in the DSP and RASSP Laboratories at Georgia Tech compare well in performance with the best VLSI implementations reported in recent literature.

### 7.1.6 Board integration

ASICs are designed to work within a system-level (board) environment. Critical issues to system integration were also highlighted and appropriate techniques for system modeling and integration were presented in this monograph. The system integration of FASICs designed within the proposed design methodology is facilitated through the use of the ***virtual prototyping*** approach introduced by Madisetti *et al* in [92] as part of ARPA's RASSP program.

The use of VHDL facilitates system portability, upgradeability, and easy integration across design tools. FASICs are exercised in the virtual system environment before commitment to hardware. Hence, design errors are caught and corrected at an early stage of the design process (when they are still easy to fix at low cost). FASICs also bridge the gap between the ASIC and FPGA approaches at the system level. This results from two facts: (1) system interface design integration of FASICs can start before hardware fabrication via virtual prototyping, and (2) the development turnaround of FASICs is dominated by fabrication time. An illustration of a board integration is given through a case-study of an i860-based board level integration. A FASIC designed within the proposed environment was integrated into an all-VHDL behavioral board model.

## 7.2 FUTURE WORK

The methodology presented in this monograph can be extended in several ways.

- **VITAL-Compliant Models:**
  In order to assure accurate system level simulation, high level VHDL models must mimic the timing behavior of synthesized circuits. When a design is resized it must be tagged to specific hardware that has its own timing characteristics given an architecture and a specific technology. The timing characteristics of resized circuits must be back-annotated to the high-level VHDL models. VITAL (VHDL Initiative Towards ASIC Libraries) [74] is the standard that defines the back-annotation process. Automatic derivation of VITAL compliant-models from the proposed library provides for a sign-off quality that can be exploited at the system validation level.

- **DSP Library Enlargement:**
  In order for the FASIC environment to encompass all aspects of DSP hardware development, the VHDL DSP library defined within the FASIC environment must be extended to include more cores and architectures that of interest to organizations. A large variation of macro-components and architectures for the same algorithm should be provided in order for the application to be added to the automatic design line.

- **Automatic Model Generators:**
  In order for the hardware design of a specific algorithm to be added to the automatic design line, a flexible array of cores that providing various different features (i.e., pipelining, multiplexing) should be designed. An efficient way would be to automatically generate different-performance (optimized for different criteria such as area, speed, and power) cores from an original basic core. A similar initiative described in [99] is aimed at automatically creating VHDL models for complex processors.

- **Automatic Verification:**
  The verification process illustrated by Figure 4.13 was performed manually. The automation of this process within the FASIC environment would add value to the proposed approach.

- **Board-level Integration:**
  Integration of new components into a board usually requires modification of the source (control, diagnostic and application) code for both the system and its components. Guidelines can be set to allow a standard encapsulation of system and components that can ultimately lead to automatic integration of systems. A similar initiative is described in [100] and also as part of Lockheed Martin ATL Laboratories' RASSP effort. The aim is to encapsulate systems and components via standard interfaces that provide a view that is independent of the internal functionality.

- **Legacy System Upgrades**
  System integration and validation in VHDL of a new core requires the development of VHDL models of all parts of the system that interact with the core. Large legacy systems exist that must be upgraded cheaply and correctly. The modeling, re-prototyping, or the upgrade of such systems is usually difficult due to the lack of documentation or models. Despite its importance, very little work has been done in recent literature. An interesting approach based on VHDL modeling is described in [101].

# A

# DWR SOFTWARE

We now describe the organization of the VHDL-based specification and synthesis environment, together with a few examples illustrating the VHDL coding style. An Automated Design Line (ADL) for FIR filters, called ADL-FIR, is utilized at one of the examples used to illustrate the methodology. At the highest level of abstraction, the VHDL architecture files reuse cores and datapath components at the lower hierarchy levels as described in earlier chapters. A detailed description of synthesis tools is available from various EDA vendors and is beyond the scope of this monograph.

## A.1 AN ADL-FIR DESIGN

We now describe the synthesis of FIR filters from high level functional specifications that include:

- data precision (wordlength),
- coefficient precision,
- internal arithmentic precision,
- number of filter coefficients.

The result of the synthesis process is an ASIC configured as a {Datapath, Controller} tuple conforming to the original design specification. The ADL-FIR branch (shown in Figure XX ) corresponds to:

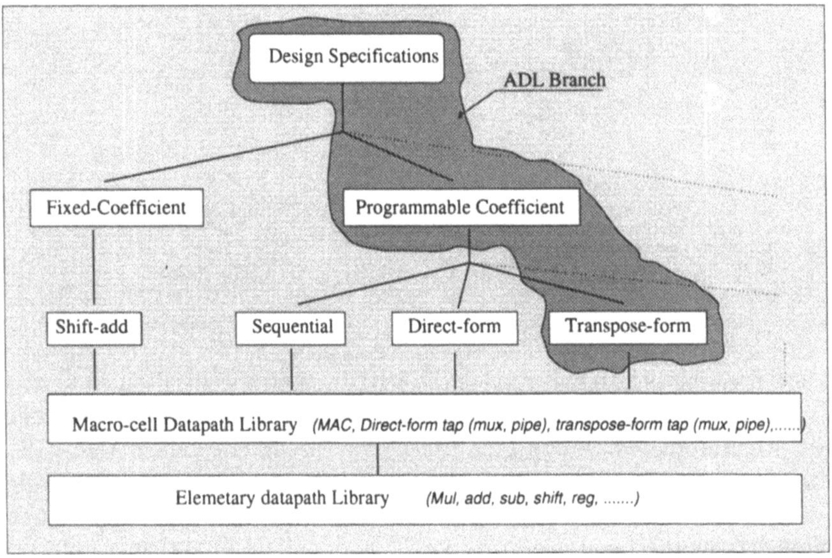


**Figure A.1** *Automatic Design Line (ADL) for FIR filters.*

$$\begin{array}{lcl} Domain & \equiv & FIR \\ Structure & \equiv & Parallel\ transpose\ form \\ Feature & \equiv & \left\{ \begin{array}{l} Double\ programmability \\ 0/2/4\ Multiplexing\ degree\ alternatives \\ 0/1/2/3\ Pipelining\ level\ alternatives \end{array} \right. \end{array}$$

Figure A.2 illustrates the flow diagram for a single branch of the ADL in the environment. A shell script is invoked via the command "executem" at the prompt. The following lists the synthesis of a 4-tap FIR filter using the core methodology within the COMPASS design environment. The specification "*top_gun.vhd*", generates the VHDL file for the controller, "*ctl_gen.vhd*", and a VHDL file for the datapath, "*filter.vhd*". Figure A.3 illustrates the relationship between ASIC VHDL files, architecture files, and cores.

```
LISTING FROM 4-TAP FIR SYNTHESIS IN COMPASS
```

```
                      ''executem''
                           |
              ---------------------------------
              User is asked to enter Parameters:
                a. data precision
                b. coefficient precision
                c. intermediate precision
                d. accumulator precision
                e. number of tabs
              ---------------------------------
                           |
 --------------------------------------------------------------
 A package : "passpara.vhd" is automatically created
 passpara.vhd passes the parameters to the architecture files
 and the macro-cell library components.
 --------------------------------------------------------------
                           |
 --------------------------------------------------------------
 Several  filter generators called ''top_gunX'' are offered to
 the designer. The designer has to choose the one that fits the
 target application, knowing the characteristics of every
 alternative. The choice is for multiplexing/pipelinig.
 --------------------------------------------------------------
                           |
 --------------------------------------------------------------
 The design tool (COMPASS Design Automation) is then invoked
 and the design is automatically generated.
 --------------------------------------------------------------
                           |
                   -----------------
                  |                 |
ctl_genX.vhd ---> CONTROLLER        DATAPATH <----- FilterX.vhd
                  |                 |
                  --------> <-------
                           |
                        NETLIST
```

**Figure A.2** *The Design Steps in the Synthesis of FIR ASICs.*

```
a. Synthesis of parallel double programmable FIR filters:
----------------------------------------------------------

              top_gun.vhd
                  |
        -------------------------------------------------
        |                         |                       |
   ctl_gen.vhd               filter.vhd             passpara.vhd
                                  |
                      -------------------------
                      |                         |
           cell1_front_end.vhd             cell1.vhd


b. Synthesis of parallel Muxed by 2 double programmable FIR filters:
---------------------------------------------------------------------

              top_gun2.vhd
                  |
        -------------------------------------------------
        |                         |                       |
   ctl_gen2.vhd                filter2.vhd              passpara.vhd
                                  |
                      -------------------------
                      |                         |
           cell5_front_end.vhd             cell5.vhd


c. Synthesis of parallel Muxed by 4 double programmable FIR filters:
---------------------------------------------------------------------

              top_gun4.vhd
                  |
        -------------------------------------------------
        |                         |                       |
   ctl_gen4.vhd                filter4.vhd              passpara.vhd
                                  |
                      -------------------------
                      |                         |
           cell9_front_end.vhd             cell9.vhd
```

**Figure A.3** *Interconnections between ASIC files, architecture files and cores.*

```
$ executem

$ Number of Taps is : 3      <- enter the number of taps here

$ Data Precision is : 9      <- enter the data precision here

$ Coefficient Precision is : 9  <- enter the coefficient precision here

$ Intermediate Precision is: 19 <- enter the intermediate precision here

$ Accumulator Precision is : 19 <- enter the accumulator precision here

$ Enter the file name (ls to list dir):

 top_gun                        <- enter the name of the model generator

$ COMPASS Design Automation Tools v8r4.4

$ COMPASS> ASICsyn--

$ Reading package PASSPARA from file PASSPARA.vhd

$ [vhd]passpara parsed with no errors or warnings

$ [vhd]top_gun parsed with no errors or warnings

$ [vhd]ctl_gen parsed with no errors or warnings

$ [vhd]filter parsed with no errors or warnings

$ [vhd]cell1_front_end parsed with no errors or warnings

$ [vhd]cell1 parsed with no errors or warnings

$ Implementing CELL1 in CELL1_p

$ Datapath library model read for datapath synthesis

$ Doing initial resource allocation for datapath CELL1_DP1

$ Optimizing the resource allocation

$ Interconnecting resources

$ Creating datapath specification

$ New datapath specification netlist is in [nls]cell1_dp1_dps

$ New datapath parameter cell is in [pcl]cell1_dp1

$ Note : [pcl] generated needs to be compiled to get the compiled-layout
```

```
$ Report file is in [lg]cell1_dp1_as
$ Doing initial resource allocation for datapath CELL1_DP2
$ Optimizing the resource allocation
$ Interconnecting resources
$ Creating datapath specification
$ New datapath specification netlist is in [nls]cell1_dp2_dps
$ New datapath parameter cell is in [pcl]cell1_dp2
$ Note : [pcl] generated needs to be compiled to get the compiled-layout
$ Report file is in [lg]cell1_dp2_as
$ Completed Datapath Synthesis Phase
$ synthesizing DATAOUT
$ Implementing CELL1_FRONT_END in CELL1_FRONT_END_p
$ Doing initial resource allocation for datapath CELL1_FRONT_END_DP1
$ Optimizing the resource allocation
$ Interconnecting resources
$ Creating datapath specification
$ New datapath specification netlist is in [nls]cell1_front_end_dp1_dps
$ New datapath parameter cell is in [pcl]cell1_front_end_dp1
$ Note : [pcl] generated needs to be compiled to get the compiled-layout
$ Report file is in [lg]cell1_front_end_dp1_as
$ Doing initial resource allocation for datapath CELL1_FRONT_END_DP2
$ Optimizing the resource allocation
$ Interconnecting resources
$ Creating datapath specification
$ New datapath specification netlist is in [nls]cell1_front_end_dp2_dps
$ New datapath parameter cell is in [pcl]cell1_front_end_dp2
```

```
$ Note : [pcl] generated needs to be compiled to get the compiled-layout

$ Report file is in [lg]cell1_front_end_dp2_as

$ Completed Datapath Synthesis Phase

$ synthesizing STORE_DATA

$ synthesizing DATAOUT

$ Implementing FILTER in FILTER_p

$ Doing initial resource allocation for datapath FILTER_DP1

$ Optimizing the resource allocation

$ Interconnecting resources

$ Creating datapath specification

$ Completed Datapath Synthesis Phase

$ synthesizing DATAOUT

$ synthesizing ACCZ_idx4

$ Implementing CTL_GEN in CTL_GEN_p

$ synthesizing W_BANK

$ synthesizing EN_OUTPUT

$ synthesizing ADD1

$ synthesizing ADD2

$ synthesizing CURRENTSTATE

$ synthesizing NEXTSTATE

$ synthesizing CURRENTSTATE_n

$ Implementing TOP_GUN in TOP_GUN_p

$ Doing initial resource allocation for datapath TOP_GUN_DP1

$ Optimizing the resource allocation

$ Interconnecting resources

$ Creating datapath specification
```

```
$ Completed Datapath Synthesis Phase
$ synthesizing DATAOUT
$ Improving hierarchy defined by TOP_GUN_p
$ setting-up parameters for the area optimization
$ Average pins per net: 2.4;  Max pins per net: 5
$ (excluding power and ground nets)
$
$Output netlist is in [nls]CTL_GEN_p
                  Num   Gate Eqv   Tot Gate     Width      Total
  Cell Name     Insts   Per Cell     Equivs  Per Cell      Width
  ---------     -----   --------   --------  --------   --------
  an02d1            1        1.5        1.5      12.8       12.8
  an03d1            2        2.3        4.5      16.0       32.0
  ao03d1            1        4.3        4.3      28.8       28.8
  dfctnb            3        7.5       22.5      67.2      201.6
  in01d0            6         .5        3.0       6.4       38.4
  nd02d0            3        1.0        3.0       9.6       28.8
  nr02d0            2        1.0        2.0       9.6       19.2
  nr03d0            6        1.5        9.0      12.8       76.8
  xn02d1            1        3.3        3.3      22.4       22.4

  Totals:          25                  53.0                460.8
  Total pins: 99
  Total nets: 35
  Average pins per net: 2.8;  Max pins per net: 8
  (excluding power, ground, and clock nets)
```

```
Pins on clock nets:

CLK                    4


Output netlist is in [nls]TOP_GUN_p

                 Num    Gate Eqv    Tot Gate       Width       Total

Cell Name      Insts    Per Cell      Equivs    Per Cell       Width

---------      -----    --------    --------    --------    --------

CTL_GEN_p          1        53.0        53.0       460.8       460.8

FILTER_p           1          .0          .0          .0          .0


Totals:            2                    53.0                   460.8

Total pins: 154

Total nets: 76

Average pins per net: 2.0;  Max pins per net: 3

(excluding power and ground nets)

DESIGN HAS BEEN SUCCESSFULLY COMPLETED !!!

A Netlist top_gun_p.nls Has been created
```

## A.1.1 Core generators

An ADL environment is composed of the invocation of a succession of VHDL core generators parameterized by high-level generics provided by the user. The top level specification is a configuration file that specifies the entire ASIC chip {Controller, Datapath, Glue logic, Memory}. In Figure A.3(b), "top_gun2.vhd" is a core generator for a $multiplexed-by-two$ double programmable FIR filter. The design parameters (i.e., precision, number of taps, etc.) for the filter are taken form the "passpara.vhd" specification file that can be modified by the user to fit the target application.

## *ASIC core generator*

The standard VHDL code for core generator "top_gun2.vhd" is listed below. The code is composed of:

1. *entity*— defines in/out ports of the chip including the interface to the target board.
2. *components*— the controller and the datapath generators are listed as components.
3. *port map*— the components and the entity are interconnected through a VHDL port map.

```
--------------------------------------------------------------------
-- Title:          ASIC generator for a 2-Muxed PARALLEL FIR FILTER
-- Filename:       top_gun2.vhd
--------------------------------------------------------------------
use WORK.passpara.all;
library IEEE;
use IEEE.STD_LOGIC_1164.ALL;
library COMPASS_LIB;
use COMPASS_LIB.COMPASS.ALL;
entity top_gun2 is
    port ( datain : in bit_vector(P downto 0);    -- Data to be filtered
           clk, reset : in bit;                   --
           work, load, switch : in bit;           -- control signals
           add : in integer range 0 to Q;  -- address to coefficient registers
           coef : in bit_vector(P downto 0);      -- coefficient bus
           acc : in bit_vector(M  downto 0);    -- accumulator input
           dataout : out bit_vector(M  downto 0) ); -- filtered output
```

```
end top_gun2;

architecture gen_top_gun2 OF top_gun2 is
signal s1, s2 : bit;
signal s3, s4 : bit_vector(Q downto 0);
component ctl_gen2 --  Controller ---------
    port  (clk, reset, work, load, switch : in bit;
           add : in integer range 0 to Q;
           w_bank  : out bit;
           sel : out bit;
           add1, add2 : out bit_vector(Q downto 0));
end component;
component filter2 -- ------- Datapath -----------
    port  (datain : in bit_vector(P downto 0);
           clk, bank, sel : in bit;
           add1, add2 : in bit_vector(Q downto 0);
           coef : in bit_vector(P downto 0);
           acc : in bit_vector(M  downto 0);
           dataout : out bit_vector(M  downto 0) );
end component;
begin
      U1: ctl_gen2 port map (clk, reset, work, load,
          switch, add, s1,s2, s3, s4);
      U2: filter2 port map (datain, clk, s1, S2, s3,
          s4, coef, acc,dataout);
end gen_top_gun2;
```

## *Core generator for the controller*

The controller core generator *"ct_gen2.vhd"* is provided below. It is composed of two processes:

1. *ctl_gen2_process*— is a combinational process with a sensitivity list containing the *CURRENTSTATE.* input signals (*work, load, switch, add,*), and *sel.* The combinational process is implemented with the VHDL *case* statement. Based on the *CURRENTSTATE* and the input control signals, the finite state machine determines the next state, and assigns it to the *NEXTSTATE* signal. The appropriate control outputs, (*w_bank, sel, add1,* and *add2*), are also generated for every specific case.

2. *ctl_gen2_seq*— is a sequential process with a sensitivity list containing the *clk,* and the *reset* signals. At each clock cycle the *NEXTSTATE* signal is assigned to the *CURRENTSTATE* signal. Based on the *NEXTSTATE* signal and the input control signals, the finite state machine ripples through all eight predefined states {*S0,S1,S2,S3a,S3b,S4a,S4b,S5*}.

```
------------------------------------------------------------------------

-- Title:       FSM GENERATOR FOR A 2-Muxed PARALLEL FIR FILTER

-- Filename:    ctl_gen2.vhd

------------------------------------------------------------------------

package DATATYPE2 is

type STATE is (S0,S1,S2,S3a,S3b,S4a,S4b,S5);

end DATATYPE2;

use WORK.passpara.all;

use WORK.DATATYPE2.all;

library IEEE;

use IEEE.STD_LOGIC_1164.ALL;

library COMPASS_LIB;

use COMPASS_LIB.COMPASS.ALL;

use COMPASS_LIB.COMPASS_ARITH.ALL;
```

```
entity ctl_gen2 is
  port  (clk, reset, work, load, switch : in bit;
         add : in integer range 0 to Q;
         w_bank  : out bit;
         sel : out bit;
         add1, add2 : out bit_vector(Q downto 0));
end ctl_gen2;
architecture behavior of ctl_gen2 is
signal CURRENTSTATE  : STATE;
signal NEXTSTATE     : STATE;
begin --------------------begin combinatorial process
ctl_gen2_process: process(CURRENTSTATE,work,load,switch,add,sel)
  begin
  case CURRENTSTATE is
     when S0 =>
        if load ='0' then
     add1 <= itobv(0, Q+1); -- initialize address register bank 1
     add2 <= itobv(0, Q+1); -- initialize address register bank 2
           sel <= '0';
           if switch = '0' then
           w_bank <= '0';      -- set working bank to register bank 1
           else
           w_bank <= '1';      -- set working bank to register bank 2
           end if;
           NEXTSTATE <= S0;
        else
```

```
    if switch = '0' then
       add1 <= itobv(2 ** add, Q+1); -- decoding address for bank 1
       add2 <= itobv(0, Q+1);
       w_bank <= '0';
       sel <= '0';
         else
        add1 <= itobv(0, Q+1);
     add2 <= itobv(2 ** add, Q+1); -- decoding address for bank 1
        w_bank <= '1';
       sel <= '0';
       end if;
       NEXTSTATE <= S1;
      end if;
      when S1  =>  ---- define the control for this sate
      when S2  =>  ---- define the control for this sate
      when S3a =>  ---- define the control for this sate
      when S3b =>  ---- define the control for this sate
      when S4a =>  ---- define the control for this sate
      when S4b =>  ---- define the control for this sate
      when S5  =>  ---- define the control for this sate
end case;
end process ctl_gen2_process;

ctl_gen2_seq : process(clk,reset)
begin  ---------------------begin sequential process
--compass stateMachine adj CURRENTSTATE
```

```
        if(To_X01(reset) = '0') then
           CURRENTSTATE <= S0;
        elsif (rising_edge(clk)) then
 CURRENTSTATE <= NEXTSTATE; -- finite state machine transition
        end if;
end process ctl_gen2_seq;
end behavior;
```

## *Core generator for the datapath*

The core generator for the datapath is listed below. The datapath is built up by a composition of two basic cores from the reusable library; *cell5_front_end*, and *cell5*. The composition of these components is realized through the combination of the VHDL *generate* statement and the parameterization of its range.

```
-----------------------------------------------------------------------
-- Title:    DATAPATH  GENERATOR FOR A 2-Muxed PARALLEL FIR FILTER
-- Filename: filter2.vhd
-----------------------------------------------------------------------
use WORK.passpara.all;
library IEEE;
use IEEE.STD_LOGIC_1164.ALL;
library COMPASS_LIB;
use COMPASS_LIB.COMPASS.ALL;

entity filter2 is
    port ( datain : in bit_vector(P downto 0);
           clk, bank, sel : in bit;
```

```
        add1, add2 : in bit_vector(Q downto 0);
        coef : in bit_vector(P downto 0);
        acc : in bit_vector(M  downto 0);
        dataout : out bit_vector(M  downto 0) );
end filter2;
architecture gen_filter2 OF filter2 is
component cell5_front_end
    port ( datain : in bit_vector(P downto 0);
         clk, bank, sel: in bit;
         regen_bank1, regen_bank2 : in bit_vector (1 downto 0);
         coef : in bit_vector(P downto 0);
         acc : in bit_vector(M  downto 0);
         store_data : inout bit_vector (P  downto 0);
         dataout : out bit_vector(M  downto 0) );
end component;
component cell5
    port ( datain : in bit_vector(P downto 0);
         clk, bank, sel: in bit;
         regen_bank1, regen_bank2 : in bit_vector (1 downto 0);
         coef : in bit_vector(P downto 0);
         acc : in bit_vector(M  downto 0);
         dataout : out bit_vector(M  downto 0) );
end component;
signal accz : REGZ;
signal s1 : DATAX;
begin
```

```
accz(0) <= acc;

G: for i in 0 to mux2_Q generate

     G1: if i = 0 generate

       cell5_front_end0 : cell5_front_end port map (datain, clk, bank,

       sel, add1(2*i + 1  downto 2*i), add2(2*i + 1  downto 2*i),

       coef, accz(i), s1, accz(i+1));

       end generate;

     G2: if (i/=0) generate

  cell5i : cell5 port map (s1, clk, bank, sel,  add1(2*i + 1 downto 2*i),

  add2(2*i+1  downto 2*i),  coef, accz(i), accz(i+1));

        end generate;

       dataout <= accz(mux2_Q + 1);

end generate;

end gen_filter2;
```

## *Macrocell model*

The core model for a multiplexed-by-two FIR tap is listed below. The core of the core is a VHDL process sensitive to the clock signal. Each declared signal is synthesized as a register. The synthesized circuit corresponds to the one illustrated in chapter 5.

```
----------------------------------------------------------------------

-- Title:        CORE: A  2-Muxed PARALLEL FIR TAP

-- Filename:     cell5.vhd

----------------------------------------------------------------------

use WORK.firpara.all;

library IEEE;
```

```
use IEEE.STD_LOGIC_1164.ALL;
library COMPASS_LIB;
use COMPASS_LIB.COMPASS.ALL;
entity cell5 is
   port ( datain : in bit_vector(P downto 0);
       clk, bank, sel: in bit;
       regen_bank1, regen_bank2 : in bit_vector (1 downto 0);
       coef : in bit_vector(P downto 0);
       acc : in bit_vector(M  downto 0);
       dataout : out bit_vector(M  downto 0) );
end cell5;
architecture behave OF cell5 is
signal REG1, REG1_bis, REG2, REG2_bis : bit_vector (P  downto 0);
signal sig1 : bit_vector (M downto 0);
begin
process(clk)
--variable declaration
----------------------------------------------------------------------
   variable mul : bit_vector  (L downto 0);
   variable tmp1, tmp2, tmp3 : bit_vector (P downto 0);
   variable mul1 : bit_vector (M downto 0);
   variable tmp : bit_vector  (M downto 0);
----------------------------------------------------------------------
   begin
   if clk'event and clk = '1' then
      if regen_bank1(0)= '1' then
```

```
REG1 <= coef;      -- This synthesizes one register for coefficient 1
   end if;
   if regen_bank2(0)= '1' then
REG1_bis <= coef; -- This synthesizes a shadow register for coefficient 1
   end if;
   if regen_bank1(1)= '1' then
REG2 <= coef;      -- This synthesizes one register for coefficient 2
   end if;
   if regen_bank2(1)= '1' then
REG2_bis <= coef; -- This synthesizes one register for coefficient 2
   end if;
   if sel = '1' then
      tmp := acc;
   else
      tmp := sig1;
   end if;
   if bank = '1' then    -- if high then  switch to the other bank
      tmp1 := REG1;
      tmp2 := REG2;
   else
      tmp1 := REG1_bis;
      tmp2 := REG2_bis;
   end if;
   if sel = '1' then   -- sel for multiplexing switch
      tmp3 := tmp1;
   else
```

```
      tmp3 := tmp2;
   end if;
   mul := datain * tmp3;
   mul1 := sign_extend(mul,M+1); -- extend to internal precision
   sig1 <= mul1 + tmp;
   if sel= '0' then
      dataout <= sig1;  -- output is passed on every  other cycle
   end if;
  end if;
end process;
end behave;
```

### A.1.2 Dataflow core generators

The following model is a Data Address Generator (DAG) model for an FFT application. The Core generator corresponds to the flow diagram shown earlier in this monograph for the FFT.

$$while\ (Count\ ¡\ M)\ do\{\cdots\}\ else\ do\{\cdots\},$$

are eliminated and replaced with dataflow type statements (i.e. *if* $\cdots$ *else*). It is suggested that there is little need in writing nested-*loop*-type code when modeling for synthesis. Instead one can write code for the core of the loop, then add control on the *start/end* execution of the core. This is often acceptable because a synthesized circuit executes at every clock cycle, hence a loop behaviour is implicit. This coding style assists in the portability of the code across VHDL compilers. However, it also requires careful post-synthesis simulation and some additional code modification (i.e., by introducing delay for signals which are used sequentially during two consecutive clock cycles; e.g., $MM \leftarrow M$ in the following code).

```
--------------------------------------------------------------------------
-- Title:         MODULE  GENERATOR FOR the DAG of the FFT
-- Filename:      dag.vhd
--------------------------------------------------------------------------
use WORK.passpara.all;
library ieee;
use ieee.std_logic_1164.all;
library compass_lib;
use compass_lib.compass.all;
entity dag is port(
clk : in bit;
rst : in bit;
done: out bit;
r1,r2,r3,r4  : out integer range 0 to N); -- N=1023
end dag;

architecture dag of dag is


signal  count  : integer range 0 to N;
signal  B      : integer range 0 to N+1;
signal  M      : integer range 1 to Q; -- Q=256
signal  MM     : integer range 1 to Q;
constant cst2  : integer:=N;
begin
process(clk,rst)
variable tmp1 : integer range 0 to N;
```

```
variable tmp2 : integer range 0 to N;
variable tmp3 : integer range 0 to N;
variable tmp4 : integer range 0 to N;
   begin
        if rst='1' then
        ----initialization:
            MM <=1;
            M <= 1;
            count <= 0;
            done <='0';
            r1 <= 0;
            r2 <= 0;
            r3 <= 0;
            r4 <= 0;
            tmp1 :=0;
            tmp2 :=0;
            tmp3 :=0;
            tmp4 :=0;
            B <= 0;
         elsif clk'event and clk = '1' then
         ----proceed address generation
             if count /= MM then
                count  <= count +1;
                tmp1   := tmp1 +1;
                tmp2   := tmp2 +1;
                tmp3   := tmp3 +1;
```

```
        tmp4   := tmp4 +1;
        r1     <= tmp1;
        r2     <= tmp2;
        r3     <= tmp3;
        r4     <= tmp4;
     else
        count <= 1;
        tmp1  := B;
        tmp2  := B + M;
        tmp3  := B + (2*M);
        tmp4  := B + (3*M);
        r1  <= tmp1;
        r2  <= tmp2;
        r3  <= tmp3;
        r4  <= tmp4;
        MM <=M;               --  MM is M(i-1)
        B      <= B + 4*M;
        if B = cst2 then
           B <=0;
           M <=4*M;
           if (M = cst2/2) then
               done <= '1';
           end if;
        end if;
      end if;
  end if;
```

```
end process;
end dag;
```

### A.1.3 RTL core generator

The simplest RTL modeling is the replication of the interconnections of an RTL schematic. The following listing illustrates the RTL VHDL model for the COMAS core of earlier chapters. The important thing when writing RTL code is to know how to control the synthesis of registers. While signals are usually synthesized into registers, one can avoid synthesizing registers for variables by ensuring that the values used are consumed quickly (before the next iteration).

```
-------------------------------------------------------------------------------
-- Title:          CORE COMAS
-- Filename:       comas.vhd
-------------------------------------------------------------------------------
use WORK.passpara.all;
library ieee;
use ieee.std_logic_1164.all;
library compass_lib;
use compass_lib.compass.all;
entity comas is
port (ar : in std_logic_vector(0 to P);
      ai : in std_logic_vector(0 to P);
      wr : in std_logic_vector(0 to Q);
      wi : in std_logic_vector(0 to Q);
```

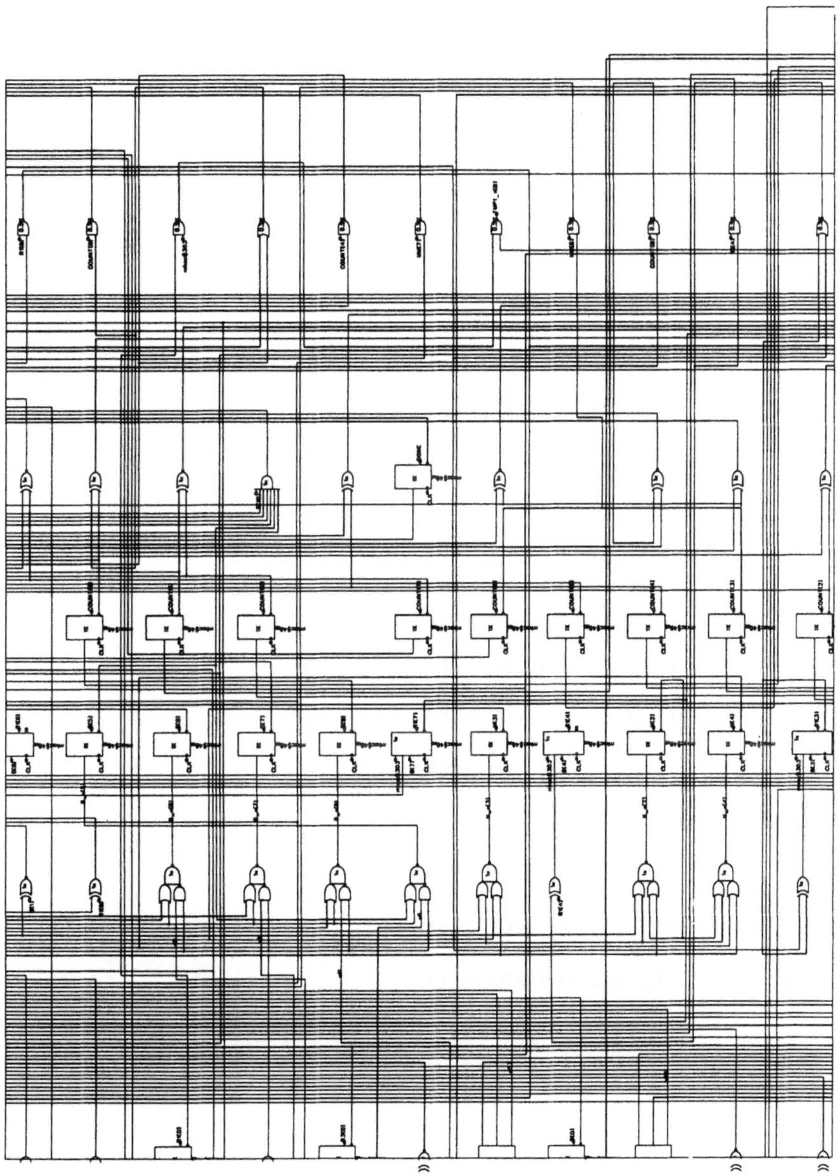

**Figure A.4** *Synthesized Dataflow in the Data Address Generator.*

```
      br : in std_logic_vector(0 to P);
      bi : in std_logic_vector(0 to P);
      clk : in bit;
      yr1 : out std_logic_vector(0 to N);
      yi1 : out std_logic_vector(0 to N);
      yr2 : out std_logic_vector(0 to N);
      yi2 : out std_logic_vector(0 to N));
end comas;
architecture comas OF comas is
signal ar1  : std_logic_vector(0 to P);
signal ai1  : std_logic_vector(0 to P);
signal wr1  : std_logic_vector(0 to Q);
signal wi1  : std_logic_vector(0 to Q);
signal br1  : std_logic_vector(0 to P);
signal bi1  : std_logic_vector(0 to P);
signal mul1 : std_logic_vector(0 to N);
signal mul2 : std_logic_vector(0 to N);
signal mul3 : std_logic_vector(0 to N);
signal mul4 : std_logic_vector(0 to N);
signal ad1  : std_logic_vector(0 to N);
signal ad2  : std_logic_vector(0 to N);
begin
process(clk)
    begin
      if clk'event and clk = '1' then
        ar1  <= ar;
```

```
        ai1  <= ai;
        wr1  <= wr;
        wi1  <= wi;
        br1  <= br;
        bi1  <= bi;
        mul1 <= ar1 * wr1;
        mul2 <= ar1 * wi1;
        mul3 <= ai1 * wr1;
        mul4 <= ai1 * wi1;
        ad1  <= mul1 - mul4;
        ad2  <= mul2 + mul3;
        yr1  <= br1 + ad1;
        yr2  <= br1 - ad1;
        yi1  <= bi1 + ad2;
        yi2  <= bi1 - ad2;
 -- note that no attempt is made to minimize area.
      end if;
end process;
end comas;
```

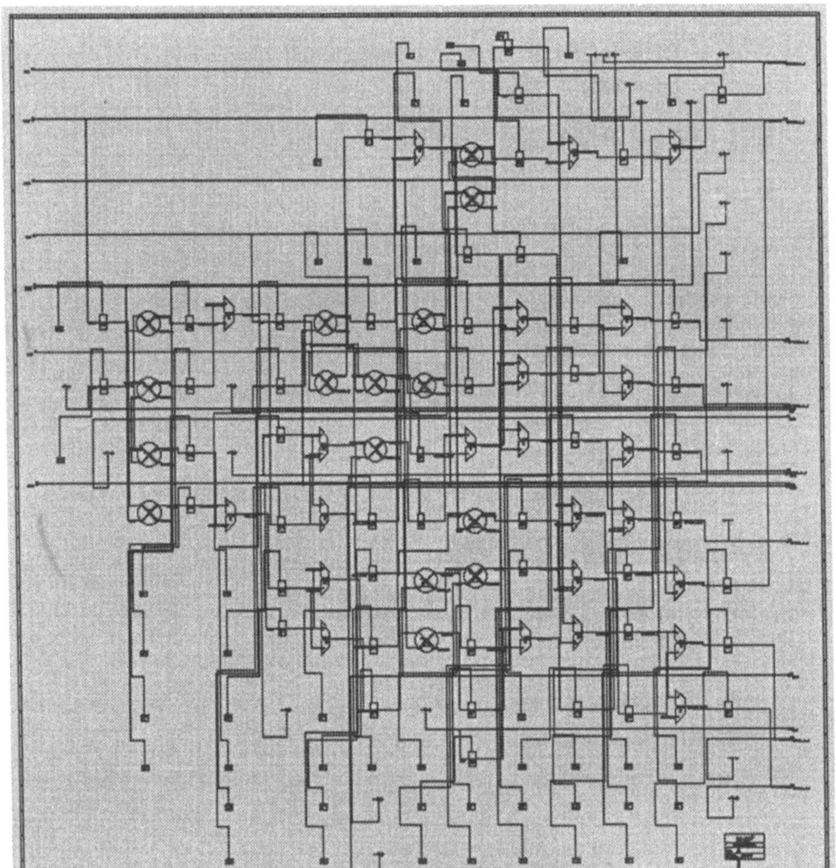

**Figure A.5** *Datapath Synthesis of the FFT datapath core (QBEU)*

# REFERENCES


[1] D. Conner, "ESDA Tools," *EDN*, pp. 80–91, May 1994.

[2] J. Fox, "A higher level of synthesis," *IEEE Spectrum*, pp. 43–47, March 1993.

[3] R. Micallef-Trigona, "Datapath intensive ASIC design - synthesis from VHDL," *IEE Colloquium on VHDL - Applications and CAD Advances*, pp. 4/1–4/5, 1993.

[4] S. Walters, "Computer-aided prototyping for ASIC-based systems," *IEEE Design and Test of Computers*, pp. 4–10, June 1991.

[5] D. Boyer and R. Cordell, "Rapid prototyping of high-speed communications chips," *IEEE Design & Test of Computers*, pp. 27–39, June 1991.

[6] H. Harvey-Horn and S. Nance, "Libraries aid complex-ASIC design," *ASIC & EDA*, pp. 52–56, March 1993.

[7] L. Maliniak, "ESDA boosts CAE technology to higher levels," *Electronic Design*, pp. 61–68, December 1993.

[8] G. De-Micheli, "Higher-level synthesis of digital circuits," *IEEE Design & Test of Computers*, pp. 6–7, October 1990.

[9] R. Camposano, "From behavior to structure: high-level synthesis," *IEEE Design and Test of Computers*, pp. 8–19, October 1990.

[10] R. Camposano and W. Rosenstiel, "Synthesizing circuits from behavioral descriptions," *IEEE Trans. CAD*, vol. 8, pp. 171–180, February 1989.

[11] J. Bhasker and H. Lee, "An optimizer for hardware synthesis," *IEEE Design & Test of Computers*, pp. 20–36, October 1990.

[12] N. Dutt and D. Gajski, "Design synthesis and silicon compilation," *IEEE Design and Test of Computers*, vol. 7, pp. 8–23, Dec 1990.

[13] M. Mahmood, B. Sharma, and C. Kingsley, "A datapath synthesizer for high-performance ASICs," *IEEE 1992 Custom Integrated Circuits Conference*, pp. 5.1.1–5.1.4, 1992.

[14] B. Sharma, M. Mahmood, and A. Ginetti, "Timing modeling of datapath layout for synthesis," *Proceeding of the 1994 International Verilog HDL Conference*, pp. 80–84, March 1994.

[15] M. Henry, "An introduction to parameterized Design with VHDL," *Integrated System Design*, pp. 60–68, August 1995.

[16] D. Gajski (Ed), *Silicon Compilation.* Mass.: Addison Wesley, 1988.

[17] D. Gajski et al, *High-Level Synthesis: Introduction to chip and system design.* Boston: Kluwer Academic, 1992.

[18] W. Ecker and M. Hofmeister, "The design cube - a model for VHDL design flow representation," *EURO-DAC '92*, pp. 752–757, 1992.

[19] V. K. Madisetti, *VLSI Digital Signal Processors: An Introduction to Rapid Prototyping and Design Synthesis.* Piscataway, NJ: IEEE Press, 1995.

[20] J. R. M. Potkonjak, "Optimizing resource utilization using transformations," *IEEE Trans. CAD of Integrated circuits and systems*, pp. 277–292, March 1994.

[21] R. Camposano, L. Saunders, and R. Tabet, "VHDL as input for high-level synthesis," *IEEE Design and Test of Computers*, pp. 43–49, March 1991.

[22] M. Taliercio, G. Foletto, and L. Licciardi, "A procedural datapath compiler for VLSI full custom applications," *IEEE 1991 Custom Integrated Circuits Conference*, 1991.

[23] COMPASS Design Automation, Inc, *ASIC synthesizer for VHDL Design User Manual.* 1992.

[24] Rowson et al., "A datapath compiler for standard cells and gate arrays," *IEEE 1992 Custom Integrated Circuits Conference*, pp. 149–152, 1987.

[25] A. Cohen and M. Shechory, "Track assignment in the pathway datapath layout assembler," *1991 IEEE International Conference on Computer-Aided Design*, pp. 102–105, 1991.

[26] D. Gajski and L. Ramachandran, "Introduction to high-level synthesis," *IEEE Design and Test of Computers*, pp. 44–54, Winter 1994.

[27] A. Dewey and A. Gadient, "VHDL motivation," *IEEE Design & Test*, pp. 12–16, 1986.

[28] J. Aylor, R. Waxman, and C. Scarratt, "VHDL-feature description and analysis," *IEEE Design & Test*, pp. 17–27, 1986.

[29] S. Collis, "The requirements for a VHDL-based custom IC, ASIC, or FPGA design process," *IEE Colloquium on VHDL - Applications and CAD Advances*, pp. 7/1–7/7, 1993.

[30] P. Williams, "Issues in integrating VHDL into the total design flow," *IEE Colloquium on VHDL - Applications and CAD Advances*, pp. 8/1–8/3, 1993.

[31] C. Harrison and K. Lam, "Extending VHDL for mixed-mode simulation," *IEE Colloquium on VHDL - Applications and CAD Advances*, pp. 9/1–9/3, 1993.

[32] A. Dewey and A. D. Geus, "VHDL: Toward a unified view of design," *IEEE Design and Test of Computers*, pp. 8–17, 1992.

[33] S. Maginot, "Evaluation criteria of HDLs: VHDL compared to Verilog," *Proc. EURO-DAC '92*, pp. 746–751, 1992.

[34] D. Skillicorn, "Techniques for compiling and executing dataflow graphs," *Information Processing 89. Proceedings of IFIP 11th World Computer Congress*, pp. 27–32, 1989.

[35] K. Kavi and A. Deshpande, "Specification of concurrent processes using a dataflow model of computation and partially ordered events," *Journal of Systems and Software*, pp. 107–120, October 1991.

[36] K. Parhi and D. Messerschmitt, "Rate-optimal fully-static multiprocessor scheduling of data-flow signal processing programs," *Proc. 1989 Int. Conf. Parallel Processing*, May 1983.

[37] B. Bhattacharyya and E. Lee, "Scheduling synchronous dataflow graphs for efficient looping," *Journal of VLSI Signal Processing*, pp. 271–288, Decembre 1993.

[38] S. Kyo, S. Okasaki, and M. Mizoguchi, "Dataflow graph optimization for dataflow architectures," *NEC Research and Development*, January 1994.

[39] C. Leiserson and J. Saxe, "Optimized synchronous circuitry by retiming," *Third Caltech Conference on VLSI*, pp. 87–116, March 1983.

[40] J. Rabaey, C. Chu, P. Hoang, and M. Potkonjak, "Fast prototyping of datapath-intensive architectures," *IEEE Design & Test of Computers*, vol. 8, pp. 40–51, June 1991.

[41] L. Lucke, A. Brown, and K. Parhi, "Unfolding and retiming for high-level DSP synthesis," *1991 IEEE International Symposium on Circuits and Systems*, June 1991.

[42] K. Parhi, "Algorithm transformations for concurrent processors," *Proc. IEEE*, vol. 77, pp. 1879–1895, December 1989.

[43] K. Van Nieuwenhove et al., "Mistral 2, a true architectural synthesis tool," *DSP & Multimedia Technology*, September 1994.

[44] P. Jha and N. Dutt, "Rapid technology projection for high-level synthesis," *Proceedings 7th Inter. Conf. VLSI Design*, pp. 155–158, 1994.

[45] R. Martin and J. Knight, "Operations Research in the high-level synthesis of integrated circuits," *Computer & Operations Research*, vol. 20, pp. 845–856, October 1993.

[46] J. Liu, "Detailed model shows FPGA's true costs," *EDN*, pp. 153–158, May 1995.

[47] D. Bhatia, "Field Programmable Gate Arrays," *IEEE Potential*, pp. 16–19, February 1994.

[48] P. Athanas and A. Abbott, "Real-time image processing on a custom computing platform," *Computer*, pp. 16–24, February 1995.

[49] R. Jain, P. Yang, and T. Yoshino, "FIRGEN: A computer-aided design system for high performance FIR filter integrated circuits," *IEEE Transactions on Signal Processing*, vol. 39, pp. 1655–1668, July 1991.

[50] J. Huisken, P. Lippens, and O. McArdle, "Design of DSP systems on silicon using the Pyramid library and design tools," *VLSI Signal Processing, III*, pp. 319–329, 1988.

[51] S. Rao, M. Hatamian, and B. Ackland, "A design environment for high performance VLSI signal processing," *Proceedings. 1990 IEEE International Conference on Computer Design*, pp. 147–152, 1990.

[52] D. Genin, P. Hilfinger, J. Rabaey, C. Scheers, and H. D. Man, "DSP specification using the Silage language," *ICASSP 90*, vol. 2, pp. 1056–1060, April 1990.

[53] Synopsys, "DesignWare Version 3.1," *Synopsys Product Family Manual*, 1994.

[54] COMPASS Design Automation, Inc., , *Datapath Compiler User Manual*. 1992.

[55] J. Tobias, "LSI/VLSI building blocks," *Computer*, pp. 83–101, August 1981.

[56] S. Powell and T. Cesear, "Rapid design and exploration of signal processing systems using a VHDL model generator based paradigm," *Proc. 2nd Annual RASSP Conference*, pp. 263–268, 1995.

[57] V. Savela, P. Jarvinen, A. Nummella, J. Keskinen, and J. Nurmi, "Utilization of a VHDL-based ASIC-realizable digital filter architecture library in DSP system design," *VHDL International Users' Forum*, pp. 6.21–6.29, 1994.

[58] O. Tanir, V. Agarwal, and P. Bhatt, "A specification-driven architectural design environment," *Computer*, pp. 26–35, June 1995.

[59] T. Maniwa, "Processor cores, megacells, and complex functions," *Integrated System Design*, pp. 42–48, July 1995.

[60] R. Kasai and T. Minami, "An overview of video coding VLSIs," *IEICE Trans. Electron.*, vol. E77-C, pp. 1920–1929, December 1994.

[61] S. Rao, M. Hatamian, M. Uyttendaele, S. Narayan, J. O'Neill, and A. Uvieghara, "A real-time P*64/MPEG video encoder chip," *1993 ISSCC Dig. of Tech. Papers*, pp. 32–33, February 1993.

[62] S. Bose, S. Purcell, and T. Chiang, "A single-chip multistandard video codec," *Proc. of CICC93*, pp. 11.4.1–11.4.4, May 1993.

[63] "IEEE STD 1076-1987," *IEEE Standard VHDL Language Reference Manual.*

[64] D. Perry, *VHDL.* New York, NY: McGraw-Hill, 1994.

[65] M. Becker and J. Diaz-Herrera, "Creating domain-specific libraries: a methodology and design guidelines," *Proc. 3rd Inter. Conference on Software reuse*, pp. 158–168, 1994.

[66] J. Poulin, "Measuring software reusability," *Proc. 3rd Inter. Conference on Software reuse*, pp. 126–138, 1994.

[67] D. Cabanis and S. Medhat, "Object-orientation applied to VHDL descriptions," *VHDL International Users' Forum*, pp. 3.9–3.15, 1995.

[68] G. Powerly Jr. and J. DeGroat, "Object-oriented generation of VHDL synthesizable architectural building blocks," *VHDL International Users' Forum*, pp. 3.1–3.7, 1995.

[69] B. Covnot, D. Hurst, and S. Swamy, "OO-VHDL: An object-oriented VHDL," *VHDL International Users' Forum*, pp. 6.1–6.10, 1994.

[70] M. S. Ben Romdhane and V. K. Madisetti, "Rapid prototyping of DSP chip-sets via functional reuse," *IEEE ICASSP-96*, 1996.

[71] G. Fettweis and H. Meyr, "High-speed parallel Viterbi decoding: algorithm and VLSI architecture," *IEEE Communications Magazine*, pp. 46–55, May 1991.

[72] G. Fettweis and H. Meyr, "High-rate Viterbi processor: A systolic array solution," *IEEE Journal on Selected Areas in Communications*, vol. 8, pp. 1520–1533, October 1990.

[73] J. Outerhout, *Tcl and the Tk Toolkit.* Addison-Wesley, 1994.

[74] "VITAL–model development specification," *VITAL–VHDL Initiative Toward ASIC Libraries Manual*, 1995.

[75] S. Lu, "An automatic model generator for VITAL-compliant ASIC libraries," *VHDL International Users' Forum*, pp. 2.1–2.8, 1994.

[76] A. Crow, "Practical VITAL 2.2b model writing," *VHDL International Users' Forum*, pp. 2.29–2.38, 1994.

[77] M. S. Ben Romdhane and V. K. Madisetti, "LMSGEN: A prototyping environment for programmable and adaptive digital filters in VLSI," *Chapter in VLSI Signal Processing-VIII, (Editor: Jan Rabaey)*, November 1994.

[78] J. Treichler, J. C. Johnson, and M. Larimore, *Theory and Design of Adaptive Filters.* J. Wiley & Sons, 1987.

[79] R. Haimi-Cohen, H. Herzberg, and Y. Be'ery, "Delayed adaptive LMS filtering: current results," *Proc. IEEE int. Conf. Acoustics, Speech, and Signal Processing*, pp. 1273–1276, April 1990.

[80] N. Shanbhag and K. Parhi, "A pipelined LMS adaptive filter architecture," *25th Asilomar Conference on Signals, Systems and Computers*, vol. 2, pp. 668–672, November 1991.

[81] G. Clark, S. Mitra, and S. Parker, "Block implementation of adaptive digital filters," *IEEE Transactions on Acoustics, Speech, and Signal Processing*, vol. assp-29, pp. 744–752, June 1981.

[82] C. Burrus, "Block implementation of digital filters," *IEEE Transactions on Circuit Theory*, vol. ct-18, pp. 697–701, November 1971.

[83] M. S. Ben Romdhane and V. K. Madisetti, "All-digital oversampled front-end sensors," *IEEE Signal Processing Letters*, February 1996.

[84] M. S. Ben Romdhane and V. K. Madisetti, "Self-correcting decision feedback equalization (SCDFE) for impulsive channels," *The Second International Workshop on: Intelligent Signal Processing and Communications Systems (ISPACS93)*, October 1993.

[85] K. Sharp, "Understanding antialiasing filters," *Electronic Design*, pp. 67–75, Dec 1990.

[86] M. Copeland, "VLSI for analog/digital communications," *IEEE Communications Magazine*, pp. 25–30, May 1991.

[87] H. Ahmed and R. Kline, "Recent advances in DSP systems," *IEEE Communications Magazine*, pp. 32–45, May 1991.

[88] R. Bhatia, M. Furuta, and J. Ponce, "A quasi-radix-16 VLSI processor," *ICASSP 91*, vol. 2, pp. 1085–1088, May 1991.

[89] J. O'Brien, J. Mather, and B. Holland, "A 200 MIPS single-chip 1K FFT processor," *ISSCC, Digest of Technical papers*, pp. 166–167, Feb 1989.

[90] H. Miyanaga and H. Y. Shroff, "A 400 MFLOPS FFT processor VLSI architecture," *IEICE Trans.*, vol. E 74, pp. 3845–51, Nov 1991.

[91] S. G. Smith, "Fourier transform machines," *Chapter in VLSI signal processing: a bit-serial approach (Editors: P. Denyer and D. Renshaw)*, 1985.

[92] V. Madisetti and T. Egolf, "Virtual prototyping of embedded microcontroller-based DSP systems," *IEEE Micro*, 1995.

[93] V. Madisetti, T. Egolf, S. Famorzadeh, and L. Dung, "Virtual prototyping of embedded DSP systems," *ICASSP-95*, pp. 2711–2714, 1995.

[94] T. Egolf, P. Kalutkiewicz, V. Madisetti, and S. Famorzadeh, "Experiences with VHDL models of COTS RISC processors in virtual prototyping for complex system synthesis," *VHDL International Users' Forum*, 1995.

[95] M. Richards, "The rapid prototyping of application specific signal processors (RASSP) program: overview and status," *Proc. 5th Intl. Workshop on Rapid System Processing*, pp. 1–6, 1994 (http://rassp.scra.org).

[96] C. Myers and R. Dreiling, "VHDL modeling for signal processor development," *ICASSP*, pp. 1–4, 1994.

[97] S. E. Shultz, "Bridging the gap to software," *ASIC & EDA*, pp. 14–25, Sep 1994.

[98] S. Famorzadeh, T. Egolf, V. Madisetti, P. Kalutkiewicz, M. Falco, and R. Dreiling, "Rapid prototyping of digital systems with COTS/ASIC components," *Proc. of the First Annual RASSP Conference*, 1994 (http://rassp.scra.org).

[99] Y. Jung, V. Madisetti, and J. Hines, "Automated generation of accurate VHDL behavioral processor models for system simulation and synthesis," *VHDL International Users' Forum*, 1995.

[100] S. Famorzadeh and V. K. Madisetti, "Rapid prototyping of DSP systems via standard interfaces," *ICASSP-96*, 1996.

[101] V. K. Madisetti and T. Egolf, "Rapid re-prototyping of legacy DSP systems," Technical Report, December 1995.

[102] A. Prabhu, "Management Perspectives in EDA," *IEEE Design & Test*, Spring 1995.

[103] V. Madisetti and B. Curtis, "A quantitative methodology for rapid prototyping and high-level synthesis of signal processing algorithms," IEEE *Trans. on Signal Processing*, Vol.42, no.11, pp. 3188-208, November 1994.

[104] J. Bhasker, *A VHDL Primer,*, Prentice-Hall, Englewood Cliffs, NJ 07632, 1994.

# INDEX

MIX
Papier aus verantwortungsvollen Quellen
Paper from responsible sources
FSC® C105338

If you have any concerns about our products,
you can contact us on
**ProductSafety@springernature.com**

In case Publisher is established outside the EU,
the EU authorized representative is:
**Springer Nature Customer Service Center GmbH**
**Europaplatz 3, 69115 Heidelberg, Germany**

Printed by Libri Plureos GmbH
in Hamburg, Germany